LANGUAGE UNLIMITED

DAVID ADGER

LANGUAGE UNLIMITED

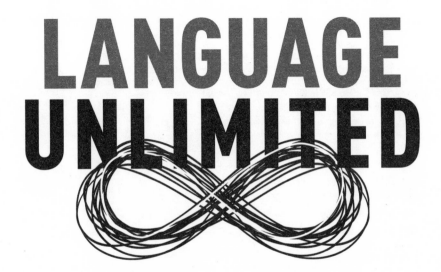

the science behind our
most creative power

OXFORD
UNIVERSITY PRESS

OXFORD
UNIVERSITY PRESS

Great Clarendon Street, Oxford, OX2 6DP,
United Kingdom

Oxford University Press is a department of the University of Oxford.
It furthers the University's objective of excellence in research, scholarship,
and education by publishing worldwide. Oxford is a registered trade mark of
Oxford University Press in the UK and in certain other countries

First Edition published in 2019

Impression: 1

Published in the United States of America by Oxford University Press
198 Madison Avenue, New York, NY 10016, United States of America

British Library Cataloguing in Publication Data
Data available

Library of Congress Control Number: 2019939545

ISBN 978-0-19-882809-9

Printed and bound in Great Britain by
Clays Ltd, Elcograf S.p.A.

CONTENTS

PREFACE

My fascination for language appeared when I was about ten years old. I'd been reading Ursula Le Guin's A *Wizard of Earthsea*, still one of my favourite books. In it, our hero Ged, is sent to a windy isolated tower on Roke, an island in the centre of Le Guin's world of Earthsea. The tower is the home of the Master Namer, Kurremkarmerruk, who teaches the core of the magical system of Earthsea: the true names of things. There, Ged learns name after name. Each plant and all its leaves, sepals, and stamens, each animal, and all their scales, feathers, and fangs. Kurremkarmerruk teaches his students that to work magic on something, you need to know the name of not just that thing, but all of its parts and their parts. To enchant the sea, Ged needed to know not just the name of the sea, but also the names of each gully and inlet, each reef and trench, each whirlpool, channel, shallows, and swell, down to the name of the foam that appears momentarily on a wave.

I found this thought fascinating even at the age of ten. I didn't really understand it, because it is paradoxical. How infinitesimal do you need to go before there are no more names? How particular do you need to be? A wave on the sea appears once, for a moment in time, and the foam on that wave is unique and fleeting. No language could have all the words to name every iota of existence. How could a language capture the numberless things and unending possibilities of the world?

I was captivated by this question. And I still am. For although Le Guin's Language of Making is mythical, human language does, in fact, have this almost mystical power. It can describe the infinite particularity of the world as we perceive it. Language doesn't do this through words, giving a unique name to each individual thing. It does it through sentences, through the power to combine words, what linguists call syntax. Syntax is where the magic happens. It takes the words we use to slice up our reality, and puts them together in infinitely varied ways. It allows me to talk about the foam I saw on a wave, the first one that tickled my bare toes on a beach in Wemyss, in Fife, on my tenth birthday. It gave Le Guin the power to put Kurremkarmerruk's Isolate Tower into the mind's eye of that same ten year old. It both captures the world as it is, and gives us the power to create new worlds.

In this book, I explain how syntax gives language its infinitely creative power. The book is a dip into the sea of the syntax of human language. It is no more than a skimming of the foam on a single wave, but I hope it gives an idea of how important understanding syntax is to the broader project of understanding human language.

David Adger

London
October 2018

1

CREATING LANGUAGE

I want to begin this book by asking you to make up a sentence. It should be more than a few words long. Make one up that, say, spans at least one line on the page. Now go to your favourite search engine and put in the sentence you've made up, in inverted commas, so that the search engine looks for an exact match. Now hit return.

Question: does your sentence exist anywhere else on the internet? I've tried this many times and each time, the answer is no. I'm guessing that that was your experience too. This isn't just a side effect of using the internet either. The British National Corpus is an online collection of texts, some from newspapers, some that have been transcribed from real conversations between people speaking English. There are over 100 million words in this collection. I took the following sentence from the corpus at random, and searched for it again, to see if it appeared elsewhere in the millions of sentences in the corpus. I then did the same on Google.

It's amazing how many people leave out one or more of those essential details.

There are no other examples. It seems crazy, but sentences almost never reoccur.

Think about your sense of familiarity with the sentences you hear or say. None of the sentences I've written so far feel new or strange. You aren't surprised when you read them. You just accept them and get on with it.

This is, if you think about it, quite remarkable. These sentences are new to you, in fact perhaps new to the human race. But they don't seem new.

The fact that sentences hardly reoccur shows us that we use our language in an incredibly rich, flexible, and creative way, while barely noticing that we are doing this. Virtually every sentence we utter is novel. New to ourselves, and, quite often, new to humanity. We come up with phrases and sentences as we need to, and we make them express what we need to express. We do this with incredible ease. We don't think about it, we just do it. We create language throughout our lives, and respond creatively to the language of others.

How can we do this? How can humans, who are finite creatures, with finite experiences, use language over such an apparently limitless range?

This book is an answer to that question. It is an explanation of what it is about human language that allows us to create sentences as we need them, and understand sentences we've never heard before.

The answer has three parts.

The first is that human languages are organized in a special way. This organization is unique, as far as we know, to humans. Sentences look as though they consist of words in a sequence, but that is not how the human mind understands them. We sense, instead, a structure in every sentence of every language. We cannot consciously perceive this structure, but it contours and limits everything we say, and much of what we think. Our sense of linguistic structure, like our other senses, channels particular aspects of our linguistic experience into our minds.

The second part of the answer is that linguistic structure builds meaning in a hierarchical way. Words cluster together and these clusters have special properties. A simple sentence, like *Lilly bit Anson*, is a complex weave of inaudible, invisible relationships. The words *bit* and *Anson* cluster together, creating a certain meaning. *Lilly* connects to that cluster, adding in a different kind of meaning. Laws of Language, universal to our species, govern the ways that this happens.

The final part of the answer tells us where this special structure comes from, and explains why we can use our languages with such flexibility and creativity. Throughout Nature, when life or matter is organized in a hierarchical way, we see smaller structures echoing the shape of the larger ones that contain them. We find this property of self-similarity everywhere. A fern frond contains within it smaller fronds, almost identical in shape, which in turn contain yet smaller ones. Lightning, when it forks from the sky, branches down to earth over and over, each new fork forming in the same way as higher forks, irrespective of scale. From slime mould to mountain ranges, from narwhale tusks to the spiraling of galaxies, Nature employs the same principle: larger shapes echo the structure of what they contain. I argue, in this book, that human language is also organized in this way. Phrases are built from smaller phrases and sentences from smaller sentences. Self-similarity immediately makes available an unending collection of structures to the speaker of a language. The infinite richness of languages is a side-effect of the simplest way Nature has of organizing hierarchies.

These three ideas, that we have a sense of linguistic structure, that that structure is governed by Laws of Language, and that it emerges through self-similarity, provide a coherent explanation of creative powers that lie at the heart of human language.

∞

I wrote this book because I think that the three core positions it takes are deep explanations of how language works. Each of these ideas is about how our minds impose structure, of a particular sort, on our experiences of reality.

Over recent years, however, an alternative to these ideas, with an impressive pedigree, has emerged. This alternative focusses not on how the mind imposes structure on our linguistic experiences, but rather on how we humans have very general powerful learning abilities that extract structure from experience.

Language, from this perspective, is like many other aspects of human culture. It is learned from our experiences, not imposed upon them by limits of the mind.

This view goes back to Darwin in his book *The Descent of Man*. The idea is that our minds are powerful processors of the information in our environment, and language is just one kind of information. The way that language works depends totally on what language users have heard or seen throughout their lives. This idea places an emphasis not on the limits of the mind, but on the organization of the world we experience.

These two different perspectives on how the mind encounters the world are both important. This book is intended to show how the first approach is better suited to language in particular.

How would language look, from a perspective where its structure emerges from our experiences?

Language, Darwin said, should be thought of in the same way as all the other mental traits. Darwin gave examples of monkeys using different calls to signify different kinds of danger, and argued that this was analogous to human language, just more limited. He argued that, since dogs may understand words like 'fetch!' and parrots might articulate 'Pretty Polly', the capacity to understand and imitate words does not distinguish us from other animals. The difference between humans and animals in language, as in everything else, is a matter of degree.

> ...the lower animals differ from man solely in his almost infinitely larger power of associating together the most diversified sounds and ideas; and this obviously depends on the high development of his mental powers.[1]

Darwin believed that humans have rich and complex language because we have highly developed, very flexible, and quite general, intellectual abilities. These allow us to pass on, augment, and refine what we do. They underpin our culture, traditions, religions, and languages. The vast range of diversity we see in culture and language is because our general mental powers are so flexible that they allow huge variation. Darwin argued that this cultural development of language augmented our ability to think and reason.

More concretely, the idea is that we can understand sentences we've never produced because we're powerful learners of patterns in general. We apply that talent to language. We hear sentences as we grow up, and we extract from these certain common themes. For example, we might hear certain words together over and over again, say, *give Mummy the toy*. We store this as a pattern, alongside *give me the banana*. As we develop, we generalize these into more abstract patterns, something like *give SOMEONE SOMETHING*, where the capitalized words stand in for lots of different things that have been heard.[2] Once this general pattern is in place, we can use it to make new sentences. The structure of our language emerges from what we experience of it as we grow up, combined with very general skills we have to create and generalize patterns. The same skills we'd use in other complex activities, like learning to bake a cake, or tie shoelaces.

Other animals have pattern matching abilities too, but, in Darwin's words, their 'mental powers' are less developed. The reason humans are the only species with syntax, from this viewpoint, is the huge gulf between us and other animals in our ability to generalize patterns. We have more oomph.

This book was written to make the argument that it's not a matter of more oomph, it's a matter of different oomph! We are not powerful pattern learners when it comes to language. We are limited—only really able use one kind of pattern for syntax, a hierarchical one. This is what I'll argue in the first half of the book. I'll also argue that patterns that depend on sequences of words are invisible to us, while syntactic hierarchy is unavailable to other animals. Though we do of course learn our languages as we grow up, what we can learn is constrained. Our limited minds are oblivious to the continuous in language, and to the sequential, and to many possible kinds of patterns that other animals can pick up on. The source of hierarchy in language is not creating patterns, storing them, and generalizing them. It's an inner sense that can't help but impose hierarchical structure, and it's the self-similarity of that structure that creates limitless sentences. That, rather than highly developed mental powers, underpins our incredible ability to use language creatively. That is our different oomph.

Unless you're an editor, or a teacher, you probably don't notice the hundreds or thousands of sentences you come across during your day. Most fly by you. In a sense you hear what they mean, without hearing what they are. But sometimes you might come across someone writing or saying something and think 'That's a bit odd.' Maybe a verb is missing. Maybe the sentence starts but doesn't end. Maybe it doesn't mean what the speaker obviously wanted it to mean. You know certain things about the sentences of your language, though you usually don't stop to think about it.

Here are some examples. Which of them are clearly sentences of English, and which are 'a bit odd'?

Zfumkxqviestblwzzulnxdsorjj kwwapotud jjqltu ykualfzgixz, zfna ngu izyqr jgnsougdd.

Sunglasses traumatize to likes that water by perplexed usually is tinnitus with amoeba an.

An amoeba with tinnitus is usually perplexed by water that likes to traumatize sunglasses.

A cat with dental disease is rarely treated by a vet who is unable to cure it.

If you're a native English speaker—and probably even if you're not—you have probably judged that the first two are not good sentences of English but the latter two are. Of these, the last one is a completely normal English sentence, while the one about the amoeba is weird, but definitely English.

If I give you many more examples of this sort, your judgments about their oddness are likely to agree with mine, and with those of many other native English speakers. Not entirely, of course. There may be words that I don't know that you do, or vice versa. Our dialects might differ in some way. I might allow *the dog needs fed*, while you might think this should be *the dog needs feeding*. You might have learned at school that prepositions are not something that we end sentences with—or not! I might not care about what they taught at school. You might be a copy editor, armed with a red pen to swiftly excise every split infinitive. I might think that split infinitives have been part of English since Chaucer, and be very happy with phrases like *to swiftly excise every split infinitive*. If we put these minor differences aside, however, we'd agree about most of it and we could agree to disagree about the rest.

How do we all do this? Why do we mostly agree?

Every speaker of every language has a store of linguistic information in their minds that allows them to create and to understand new sentences. Part of that store is a kind of mental dictionary. It grows over our lives, and sometimes shrinks as we forget words. It is a finite list of the basic bits of our language. But that's not enough. We also need something that will allow us to combine words to express ourselves, and to understand those combinations when we hear them.

Linguists call this the mental grammar. It is what is responsible for distinguishing between the first two examples and the latter two. As every speaker grows up, they learn words, but they also develop an ability that allows them to put words together to make sentences of their languages, to understand sentences, and to judge whether certain sentences are unremarkable or odd.

But do we really need a mental grammar? Maybe all we need is the mental dictionary, and we just put words together and figure out the meanings from there. Knowing what the words mean isn't, however, enough. The meanings of sentences depend on more than just the meanings of the words in them. Take a simple example like the following:

The flea bit the woman.

Using exactly the same words we can come up with a quite different meaning.

The woman bit the flea.

How we put words together matters for what a sentence means. Just knowing the meanings of words isn't sufficient. There's something more going on.

These two sentences also show us that how likely one word is to follow another makes no difference to whether we judge a sentence to be English or not. A bit of quick Googling gives about a million results for the phrase 'bit the woman' and just eight results for 'bit the flea'. This makes complete sense of course. We talk more about people being bitten than fleas being bitten. But the likelihood of these two sentences makes no difference as to whether they are both English or not. One is more probable than the other, but they are both perfect English.

The mental grammar can't be reduced to the mental dictionary plus meaning, or frequency. We need both the mental dictionary and the mental grammar to explain how each of us speaks and understands our language(s).

The question of whether we have mental grammars or not isn't really disputed. Whether we think of the human capacity for syntax as emerging from the structure of experience, or from the particular limits of our minds, we still need to say that the general rules of our particular languages are somehow stored in our minds.

But we can use the nature of what our mental grammars must be like to begin to dig into the question of the source of syntax. Is it part of our nature as human beings, or is it something we pick up from the world we experience?

∞

To ask a certain kind of question in English, you use a word like *what, who, where, when*. Take a scenario where someone is chatting away and mentions that my cat, Lilly, had caught something in the garden. I didn't quite hear the full details, so I ask:

What did you say that Lilly had caught?

Here, the word *what* is asking a question about the thing that Lilly caught. Although *what* is pronounced at the start of the sentence, it is really meant at the end. After all, we say *Lilly caught something*.

In many other languages, like Mandarin Chinese, Japanese, or Hindi, to ask a question like this you'd just leave the word for *what* right next to the word for *caught*, giving the equivalent of *You say Lilly caught what?* Here's how this looks in Mandarin Chinese:

Nǐ	shuō	Lìlì	zhuā	shénme
You	say	Lilly	catch	what

The Chinese word *shénme* corresponds to English *what*, and it comes after the verb *zhuā*, which means catch. That's the normal order of words in a Chinese sentence. In a question, nothing changes.

Let's think about how to capture this difference if what we have learned of our language, our mental grammars, develops through noticing and storing patterns from our experiences. Imagine a person, Pat, whose mental grammar grows and is refined over time in this way. Pat learns through noticing, and storing, patterns.

If Pat grew up speaking Mandarin Chinese, they would learn to treat question words no different from non-question words. If exposed to English, they would learn that a question word is placed at the start of the sentence. Pat's mental grammar in this latter case would contain a statement something like this:

If you want to ask a question about a thing, a time, a place etc., use a word like what, when, where, *etc. or a phrase like* which X, *and place this at the start of the sentence.*

Pat doesn't consciously know this, but something about Pat's mind makes them behave according to this pattern. Pat has unconsciously learned how to make and understand certain questions in English.

The word *that*, as we just saw in the example above, is used in English after words like *say*, *think*, *believe*, and so on, to introduce what is said, thought, or believed. When *that* introduces a sentence in this way, it is often optional in English. We see this in sentences like the following:

Anita said Lilly had caught a mouse.

*Anita said **that** Lilly had caught a mouse.*

We can put the word *that* in here, or leave it out. Both sentences are perfectly fine ways to express what we mean here.

It's not at all surprising, then, that we can leave out the word *that* when we ask a question too. Both of these next examples are perfectly fine ways of asking the same question:

What did Anita say Lilly had caught?

*What did Anita say **that** Lilly had caught?*

How would Pat's mental grammar look if they were an English speaker? They would have learned that the word *that* is optional after the verb *say*, and other verbs like it, so their mental grammar would contain something like this generalization:

Optionally put the word that *after verbs like* say, believe, think ...

So far, so good. Pat's mental grammar contains these two patterns, and many more.

But now let's imagine I had had a different conversation. Imagine the discussion was about one of the neighbourhood cats catching a frog in my garden. If I want to identify the cat, I can ask:

Which cat did Anita say had caught a frog?

A superficial difference between these two questions is whether we are asking about what was being caught, or who did the catching.

Given that the word *that* is optional after *say*, we expect Pat to think that the following sentence should also be fine:

Which cat did Anita say **that** *had caught a frog?*

For most speakers of English, though, this sentence is 'a bit odd'. It is much better without the *that*.

This poses a problem for Pat. They would be led to the wrong conclusion about this sentence. It is a question, using *which cat*, and, as expected, *which cat* occurs at the start. Pat, as we know, has learned a pattern which allows the word *that* to appear as an option after *say*. The sentence matches the pattern: we have taken the option to put in the word *that*. The trouble is that Pat, who is a good pattern learner, would think this sentence is perfectly fine. But most speakers of English think it's not fine—it's decidedly odd. This suggests that most speakers of English, unlike Pat, are not good pattern learners.

This argument doesn't prove that the pattern learning approach is wrong. Real English speakers could be more sophisticated than Pat is.

For example, it could be that children learning English do learn patterns like Pat does, and use those patterns to predict what they will hear. They expect to hear sentences like *What did you say* **that** *Lilly caught?*. But, the explanation goes, they never do. This means that what they experience doesn't match up with their expectations. The way that the children deal with this is to store an exception to the pattern they have learned. In this scenario, the children's experiences would contain enough structure to help them come to a more complex pattern.

This is an interesting idea, which we can test. In 2013, two linguists, Lisa Pearl and Jon Sprouse, did a careful study of the speech directed at young children who are acquiring English. They looked at over 11,000 real examples where parents, or other caregivers, speak to their children.[3]

They found that parents, when they asked their children these kinds of questions, almost always dropped the word *that*. They did this whether they were asking a question about what had had something done to it, or what was doing something. It made no difference. The parents never took the option to put *that* after words like *say, believe,* etc. This means that the children didn't ever get the information they would need to learn that there was a difference between the two types of questions.

If we think about this from Pat's perspective, the syntax of English is completely mysterious. Pat's mental grammar consists of patterns they've learned from their experiences. If Pearl and Sprouse are right, Pat couldn't have learned the exception to the pattern that allows *that* to disappear. Pat's experiences, which we are assuming are just the experiences children learning English have, aren't rich enough to learn an exception to the generalization about when *that* appears. Adult English speakers' mental grammars, however, clearly have that exception in them. This seems like a strong argument that English speakers don't work like our imaginary friend Pat. They aren't simply good pattern learners.

Intriguingly, many other languages behave in the same way as English, even though these languages are not related to English or to each other. For example, Jason Kandybowicz studied the Nupe language, spoken in Nigeria, and found exactly the same pattern there. Here's how you say *What did Gana say that Musa cooked?* in Nupe, with a word by word translation:[4]

Ké	*Gana*	*gàn*	*gànán*	*Musa*	*du*	*o?*
What	*Gana*	*say*	*that*	*Musa*	*cook*	*o*

The order of words here is quite similar to English. The little word *o* at the end marks that a question is being asked and the word *gànán* is the equivalent of English *that*. Just as in English, it is impossible to say the equivalent of *Who did Gana say that cooked the meat?* You can put the words together, but Nupe speakers don't judge it to be a sentence of Nupe:

Zě	Gana	gàn	gànán	du	nakàn	o?
Who	Gana	say	that	cook	meat	o

There are many other languages that work similarly (Russian, Wolof, French, Arabic, and some Mayan languages).[5]

It is a fascinating puzzle. Speakers end up with judgments about sentences of the languages they speak that don't depend on what they have heard as children. Certain ways of putting words together just aren't right, even though, logically, they should be. And these quite subtle patterns appear in unrelated languages over and over again. We humans seem to be biased against our languages working in perfectly reasonable ways!

There are many puzzles just like this in the syntax of human languages. Languages do have a logic, but that logic is not one that emerges from the patterns of language we experience. The linguist's task is to understand the special logic of language, what laws govern it, and how different languages find different ways to obey those laws. We'll find out in the rest of the book that it's the hierarchical structures that underlie sentences that are responsible for many of these quirks. Some are, without doubt, learned from experience, but others, as we've just seen, are not.

∞

Syntax is a deep source of human creativity. You constantly come across sentences that you've never heard before, but you have no trouble understanding them. My favourite headline of

2017 simply said *Deep in the belly of a gigantic fibreglass triceratops, eight rare bats have made a home.* Beautiful, crazy, and true. Syntax gives us the capacity to describe even the weirdest aspects of our existence, and, of course, allows us to create new worlds of the imagination.

The most basic units of language, words and parts of words, are limited. We can create new ones on the fly, if we need to, but we don't have a distinct word for every aspect of our existence, unlike the wizards of Earthsea. The number of words speakers know is a finite store, a kind of dictionary. We can add words to that store, and we can forget words. But the sentences we can create, or understand, are unlimited in number. There is no store of them.

This book makes the argument that hierarchy and self-similarity underlie our creative use of language. On the way, we'll find out why language is not just communication, how we can sense linguistic structure without being aware of it, and how sentences are like gestures in the mind. We'll meet children who cannot experience the language spoken around them, and so they create new languages for themselves, languages that are taken up by communities and become fully-fledged ways of expressing thoughts. We'll see how human languages follow particular, limited, patterns; how scientists have invented languages that break these; and how they have used these languages to test the limits of the human brain. We'll invent languages to be spoken by imaginary beings, and imagine languages that could never be used. I'll show you how rats can pick up on linguistic structures humans cannot perceive, and how humans can discern ones invisible to our closest evolutionary cousins, the apes. I'll reveal the mysteries of how AIs understand sentences, and how different that is from what we do when we speak and understand language.

We'll also do a little linguistics. You'll learn about some of the Laws that limit how human languages work, and why these Laws can be Universal without being universal. You'll also meet some unusual languages, from Chechen to Gaelic, Korean to Passamaquoddy, and Yoruba to Zinacantán Sign Language. I'll gently introduce you to one of the most cutting-edge ideas in linguistics: Noam Chomsky's proposal that one linguistic rule creates all the innumerable structures of human language. This idea provides a foundation for understanding what underlies our ability to use language in the creative ways we do, but it also leaves open a space for understanding how that use is affected by our social nature, our identity, emotions, and personal style.

2

BEYOND SYMBOLS
AND SIGNALS

In 2011, an internet entrepreneur, Fred Benenson, crowd-sourced a translation of *Moby Dick* into emojis. The word *Emoji* comes from two Japanese words: *e*, meaning picture, and *moji*, meaning a written symbol, like a Chinese character, a hieroglyph, or even a letter of the alphabet. Emojis, then, are intended to be similar to written words: they convey meaning through a written form. Because emojis seem like words, people have talked about their use as the 'fastest growing language'. The initial set of about 180 emojis has grown to over 3,000. Over five billion emojis are used every day on Facebook.

Even more exciting is the idea that emojis are somehow universal. They are pictures, so we can understand them no matter what language we speak. But they are also like words, opening up the idea that emojis could be a universal way of communicating, a language for everyone.

Are emojis like words? When we string them together in our electronic communication is that a universal language?

The linguists Gretchen McCulloch and Lauren Gawne have argued that emojis, as we actually use them, are far more like gestures than like words. They are a body language for our bod-iless internet selves. The thumbs-up, middle finger, or eye-roll emojis directly represent gestures, but the way we use other emo-jis is also gesture-like. McCulloch points out that we often repeat gestures three or four times to emphasize what we're saying, adding to speech by thumping a fist repeatedly on a table, or opening up our hands, entreatingly, in front of our bodies. The most common sequences of emojis are just repetitions: lots of smiley faces, love hearts, or thumbs-ups. We don't repeat most words in the same way—words have a place in our sentences, and few of them can be repeated without something going wrong.[1]

When we play charades, or watch mime artists, we're using and understanding a kind of pantomime. This, McCulloch argues, is very similar to the ways that you can use strings of emojis to tell stories. This is why Benenson's project was never going to work. It's the equivalent of miming the whole of *Moby Dick*.

Emojis are not really like words then. Though we use them to communicate, that communication is more like what happens with body language.

It's interesting to think about what we'd have to do to emojis to make them work more like words. Perhaps if we enriched emojis, they could work more like a universal language?

Unlike words in spoken or written language, emojis don't express sounds. Expressing sounds, though, is important for even something so simple as someone's name. Benenson's translations of characters from *Moby Dick*, like Ishmael or Queeqeg, are impenetrable. Ishmael becomes a boat, a whale, and an ok sign, signifying what roles he plays in the novel, not the sound of his name.

It is possible, though, to develop emojis so that they could express sounds. For example, you could associate certain emojis with the sounds of the English words that those emojis make you think of. An emoji for a cat could be used for the syllable *cat*, so you could express *catatonic*, say, by using a cat emoji and an emoji of a gin and tonic. Each emoji would stand for a sound, rather than for what it pictures. This would allow us to express the sounds of names. My name, for example, could be a picture of a sun rising (day) and an old style video cassette (vid).

The alphabetic system that English uses connects written letters to sounds, so it can easily represent how names are pronounced. The Chinese writing system works differently, and is similar to the original intent of emojis. It involves symbols for particular words as opposed to sounds. Because of this, it also faces challenges representing names, especially those that are not native Chinese. However, the users of this system have developed sophisticated ways of writing foreign names by using Chinese characters that have sounds similar to the syllables of the name. I was once given the Chinese name Ai Dao Fu. Surnames in Chinese come first, and usually consist of just one syllable. The Mandarin Chinese word *ài*, which means 'love', is close in sound to the first syllable of my surname (the 'a' in Adger). The words *dào* (meaning 'way', as in Daoism), and *fú* 'happiness' are close in sound, when put together, to David. Chinese has characters for the words 'love', 'way', and 'happiness', so you can use these characters with their associated sounds to write something that is pronounced a bit like my name: Ai Dao Fu—with some lovely meanings to go with the sounds.

A bit more abstract than this would be to use the cat emoji as a kind of shorthand for the sound *k*—often written in English as a *c*—that appears at the start of the word. Doing this connects the symbol to a sound and that's how many of us learned the alphabet. 'A' is for apple, 'B' is for book, 'C' is for cat, and 'D' is for dog. This basic idea has appeared again and again in the history of writing systems. Pictures which are initially used to represent ideas end up being used to represent sounds.

Ancient Egyptian hieroglyphs worked like this. The word for 'mouth' in that language was pronounced something like *re*, and it could be written using a picture of a mouth:

This hieroglyph is actually usually used to convey the sound *r*. For example, the Ancient Egyptian god Ra, the sun god, was written as the sound *r* above another hieroglyph that was used for a sound that comes out a bit like what happens when you try to cough and swallow at the same time—linguists write this, in the international phonetic alphabet like this ʕ, and the Ancient Egyptian word for 'arm' started with it:

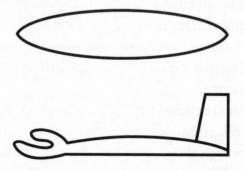

Adopting this idea would allow us to use emojis to write sounds. We could use a cat emoji for the *k* sound, an arm emoji for an *a* sound, and a cup of tea emoji for a *t* sound. We could then express the word for a furry purring animal as follows.[2]

It's rather hard to see how this would be an improvement on just texting though!

Much of the early hype around emojis was about how they were universal. Anyone who spoke any language would be able to understand them. This is certainly an exciting idea, but no symbol is truly universal to humankind.

A symbol is just some kind of a mark made on the world that stands in for something else, usually an idea in your head. This means that there are two parts to a symbol. There's a concept, something inside your mind. This is the meaning of the symbol. There's also something that is external to you, something which you can see (like an emoji), hear (like a spoken word), or feel (like Braille letters). This is called the form of the symbol. So a symbol is a connection between a mind-internal meaning and a mind-external form, between something abstract, and something concrete.

When the mind external part of the symbol, the part you can see, hear, or feel, resembles the symbol's content, then there's a direct psychological link between the two. In this case, the symbol is a bit like a computer icon, say one for a wastepaper basket. Symbols like this are called iconic symbols. Many emojis are iconic, like the ones we just saw for cat, arm, and tea.

There are also symbols without that direct link of resemblance—the relationship between the content and its expression can be quite abstract or even arbitrary. A love heart is like this. A love heart doesn't look much like a real heart, and the association between the emotion of love and an internal organ is, at best, indirect.

Could we build a universal language built out of iconic symbols like emojis? Since they'd be iconic, people should understand them regardless of what language they speak or culture they come from.

In the early 1990s the US Government commissioned a report on nuclear waste. It had the rather dry title 'Expert judgment on markers to deter inadvertent human intrusion into the waste isolation pilot plant'. A team of experts was set up to figure out how to communicate to unknown people in the far distant future, that a particular plot of land in New Mexico was going to be dangerously radioactive for many millennia. The standard symbol for nuclear waste may not be recognizable in millennia. There may be many radical environmental or cultural changes for humanity.

Various ideas were considered as possible symbols, including 'menacing earthworks', 'forbidding blocks', 'horrifying facial icons' like Munch's The Scream. Carl Sagan, the astronomer, physicist, and novelist suggested a skull and crossbones. That didn't fly. The team reported that 'The lineage of the skull and crossbones ... leads back to medieval alchemists, for whom the skull represented Adam's skull and the crossed bones the cross that promised resurrection. It is almost certainly a Western cultural artefact'. The fundamental problem is that all of the symbols of danger that the team could come up with simply might not mean danger to an unknown population in the future. As the report says,

> No symbol is certain to stay in use for the 10,000 year period. Future societies will probably create many of their own

symbols, and symbols from our time may have their meanings changed or distorted with the passage of time. Compare how the meaning of the swastika has changed in our own century, going from positive religious symbol of India to a hated emblem of the Nazis.[3]

The basic idea that emojis could be truly universal, then, could never get off the ground. Human symbols are always, in the end, deeply connected to our cultures.

Words are the crème de la crème of arbitrary symbols. Aside from a few cases, like animal noises—did you know that the Mandarin Chinese word for 'cat' is *māo?*—they are associated with their meanings through a socially agreed convention. They don't resemble them in any psychological way. This is why the word for 'dog' is *dog* in English, *txakur* in Basque, and *inja* in Xhosa—the same concept expressed by quite different sounds.

We could, then, just as we do with words of spoken languages, or the signs in sign languages, link emojis to meaning using social conventions. The resemblance relationship between an emoji and its meaning would then be useful in guessing a meaning, but the meaning itself would be fixed by communities of emoji users. In fact, such conventions have arisen already through internet users interacting with each other. Sanjaya Wijeratne, while researching his PhD at Wright State University, discovered that gang members were using a gas-pump emoji in their tweets to signify marijuana. Other researchers have found that the meaning of emojis changes across cultures. In some cultures the handwave emoji is just a sign off, in others it's a snub.[4]

Emojis then could be developed to work more like words, though, if McCulloch's gesture idea is right, it's intriguing that that has not been what has happened naturally. Such an emoji language wouldn't, however, be universal.

There is more to a language than just words, though.

If someone texted you the stream of emojis you see here, what would it mean?

Does it mean a cat is kicking something? Or someone is kicking a cat? Or is it about the story of *Puss in Boots*? Or maybe your friend wants you to get a pair of boots with a cat on them? And how would you even go about clarifying which of these you meant? In spoken or written English (or Cantonese, or Swahili), it's easy to express what you mean with a fair level of precision—in fact, I just did. When you are using emojis, the context might make the message clear. Perhaps you've already been talking about one of these topics with your friend. But in the absence of context, emojis are far too vague to work like a language.

Or think of this the other way around, in terms of expressing yourself, rather than understanding what someone else is trying to convey. How would you express, in emojis, that something has happened in the past? Or the thought that, if something were to happen, so would something else? Or that something didn't happen? Or how would you express that every cat was kicked, not lots of cats, every cat. These concepts, so easily expressed in a few words using a language like English or any other human language, are completely beyond the capacity of emojis, at least without changing what emojis are: a simple connection of a picture and an idea, obvious to everyone when they see it.

The failure of emojis to express past time, events not happening, possibilities of events taking place, quantifying objects, and hundreds of other purely grammatical ideas, gives us a clue to why emojis are different from a natural human language, even if we let emojis include arbitrary symbols. Emojis do communicate ideas using symbols, but human language goes beyond symbols and, as we will see, beyond communication.

To see this, let's go back to our cat and boot emojis. Is the cat kicking or walking? Or is it being kicked? Let's add one more emoji:

We've got a cat, a boot, and a boy. What message is being expressed?

If you speak a language like English, you might be tempted to assume that the order of the emojis is linked to the order of the corresponding English words 'cat' 'kick' 'boy' in the sentence *The cat kicked the boy*. This would give you the meaning that the cat kicked the boy. But isn't it more likely that the boy kicked the cat—after all, boys wear footwear, but cats generally don't? That would be a more sensible and likely message, so maybe you should ignore the order and just go for what is the most probable message that's being communicated.

But maybe you speak a language like Malagasy. The order of words in Malagasy is quite different to that in English. In Malagasy you'd say something like 'kicked the cat the boy', to express that the boy kicked the cat. The person doing the kicking comes last in the Malagasy sentence. This might tempt a speaker of Malagasy towards the meaning that the boy did the kicking. Or since in Malagasy the verb actually comes first, maybe you'd think that these emojis mean that the boy miaowed at the boot—maybe he was pretending to be a cat.

This discussion tells us something important: human languages have ways and means of expressing certain ideas—who did what to whom, for example—that go beyond iconic symbols. English can express who does what to whom partly by the order of the words it uses. Malagasy does the same, but uses a different kind of link between aspects of meaning and the order of the words.

Emojis, even if we enrich them, and make them true symbols, don't have this property. There's no convention about how emojis express who did what to whom.

Let's imagine we can somehow add such a convention. Let's say that the first emoji is always the individual performing some kind of action, the second emoji represents that action, and the last emoji is an individual who gets affected by the action. This is similar to what we just saw in English. Would this bring us closer to how human languages actually work? We'd be adding in a new kind of convention, a kind of extended symbol. It would still be a link between meaning and form. The meaning of who did what to whom is linked to the form, the observable order that the emojis come in.

This idea of an extended symbol doesn't really work in the way spoken or signed languages do, though. Take the simple English sentence:

The boy was kicked by the cat.

In this sentence, *the boy* comes first, but he's not doing the kicking. So although English can express who does what to whom through one particular order, it actually has many possible orders. Emojis don't lend themselves to this kind of complexity.

Another kind of interesting example is a sentence like:

The boot filled with water.

In this sentence, it's the boot that is affected by the action, and the water that's causing that filling up to happen, even though the words *the boot* come before the word *water*. This time it's the particular meaning of the verb that overrides the usual conventions.

These kinds of examples tell us that the link between the form of a sentence in a human language and what it means is quite subtle and indirect. We can't make emojis into a language by just adding in some conventions about how meanings link to orders. Languages are far more sophisticated and intricate than that.

The extended symbol idea also falls foul of one of the most important properties of the sentences of human languages: words cluster together in groups and languages are exquisitely sensitive to this grouping.

To see this, imagine I say to you that Anson is off to run a marathon up and down the mountains of Glen Coe in Scotland. You might say to your friend, Anson's doctor:

Wow! Can Anson run a marathon with his sprained calf?

Your friend, if she likes, could reply:

Yes. Anson can run a marathon with his sprained calf.

If we compare these two sentences, you can see that the difference is where the word *can* appears. If *can* appears before *Anson*, then the meaning is a question, not a statement.

Let's try to understand this difference in terms of symbols. It would again be a kind of extended symbol. Putting the word *can*, and other words like it, before *Anson* links to the meaning that a question is being asked. Putting it after, links to the meaning that a statement is being made. The position of the word *can* in the sentence is the form, linked to the question or statement meaning.

We need to be a bit more precise about the position of the word *can*. First, it's not just this word that has this effect. We can see this by looking at other similar cases. In these examples, I've put the word that shifts around in bold:

*Lilly **is** jumping.*　　　　　　**Is** *Lilly jumping?*

*The cat **has** caught a frog.*　　　　**Has** *the cat caught a frog?*

*We **did**, in fact, arrive early.*　　　**Did** *we, in fact, arrive early?*

There's a particular set of words that shifts around like this in English. They are called auxiliary verbs. We can see, in each of the statements, that the auxiliary verb appears after a certain word or phrase. In the corresponding question, it appears before that word or phrase.

This is quite abstract but could serve as the form to which the meaning is linked. At first glance, then, it looks like we can understand these statement-question examples in terms of a kind of extended symbol. The form is the order of words, as opposed to just how particular words are pronounced, and the meaning is what the form can be used for, a statement or a question. When we look a little deeper, though, we see that we need to go beyond symbols to really understand what is going on in these examples.

Since a symbol is a link between form and meaning, we'd expect that whenever we see the form, we get the meaning, and, whenever we want the meaning, we use the form. For these statement-question examples, and many others, it turns out that you can get the meaning without the form, and the form without the meaning, undermining the idea that this should be thought of symbolically.

I can express a question without putting the word *can* before *Anson*. A verb like *ask* explicitly calls for a question, as in the following sentence:

*I'll ask **if Anson can run a marathon with his sprained calf***.

What comes after *ask*, which is in bold, expresses a question, in fact the same question that is expressed by saying *Can Anson run a marathon with his sprained calf?* But the word *can* stays put. Instead, we find the word *if* at the start of the question.

Maybe there are two different extended symbols for questions then? Either we put *can* before *Anson*, or we leave it where it is and put the word *if* before *Anson*.

But that isn't sufficient. We don't, for example, just put *if* at the start of a sentence in English to make a question. Otherwise the next sentence would be a perfectly good way to ask a question in English, and it's not, though some languages do actually work like this, Scottish Gaelic, for example:

If Anson can run a marathon with his sprained calf?

Similarly, in many people's English—although not everyone's— you can't swap *can* and *Anson* around after the word *ask*. The next sentence isn't a way of saying the same thing as *I'll ask if Anson can run a marathon with his sprained calf*. It means something quite different, and would have a different punctuation:

I'll ask can Anson run a marathon with his sprained calf.

This little discussion shows that you can have the same meaning with a different form.

There are also problems for the extended symbol idea the other way around. For example, there are examples where we swap around the order of *can* and *Anson* but we don't get a question. Instead we get an even stronger statement. Our doctor friend, who may have been administering a miracle cure to Anson, could reply to our very first question like this:

Boy **can** *Anson run a marathon with his sprained calf!*

This shows us that there are different forms linked to the same meaning (two ways of making a question), and the same form linking to different meanings (two meanings swapping round *can* and *Anson*). The link between form and meaning in a language like English just isn't the same as that between form and meaning in a symbol like an emoji.

There's one final way in which these kinds of sentence show us that human languages go beyond the symbolic. A symbol, as we've seen, is a link between a concept or idea and something we can see or hear. But it turns out that, in human languages, sometimes the form of sentences is actually invisible. This means that symbols, however extended or elaborated, are just insufficient as an explanation of language.

In the following sentence the word *can* appears twice:

The person who **can** *run fastest* **can** *win the marathon.*

Now, if we want to make a question of such a statement, we say:

Can *the person who* **can** *run fastest win the marathon?*

Weirdly, we've taken the second *can* and put it at the start of the sentence, not the first one.

Maybe it's always the last *can* that is affected by the rule that makes questions? That would explain what happens in the next sentences:

*The person who **can** catch the cat that **can** run fastest **can** win the marathon.*

***Can** the person who **can** catch the cat that **can** run fastest win the marathon?*

But no. It's not the last one:

*That person **can** win any marathon you **can**.*

***Can** that person win any marathon you **can**?*

Now it's the first *can* that is placed at the start of the question.

What is happening here? What is the rule of English that picks out the right *can* in these sentences?

The best answer we have to this goes beyond the idea of symbol entirely. Think about the collection of words that *can* hops over to turn a statement into a question. We can replace these words by a single word—in this case the word *he*, given that Anson is male. We can do this no matter how long that collection of words is. I've put them in bold here so it's easy to see:

The person who can run fastest *can win the marathon.*

The person who can catch the cat that can run fastest *can win the marathon.*

He *can win the marathon.*

This replacement preserves the basic message that the sentence communicates as long as we know who *he* is being used to refer to.

This shows us that these words behave as a single group. The auxiliary verb that appears after that group in a statement, appears before it in a question. But there's nothing that visibly signals the 'groupiness' of the group. It's an invisible, inaudible, property of those words that they group together.

There are lots of other properties that single out this same group of words. But these properties are not symbolic. They don't involve a simple relationship between something you can directly perceive (a sound, or a written symbol) and a concept or meaning.

Many people will recognize that rule of English that is at work here involves the notion of a grammatical Subject. But what exactly is a Subject in English?

This is actually a pretty hard question, but here are some things it's not. It's not the first word or phrase in a sentence. *In fact* is not the Subject of:

> *In fact, Lilly will scratch the sleepy girl.*

We can see this if we try to make this sentence into a question. The word *will* hops in front of just the word *Lilly*, not *in fact*:

> *In fact, will **Lilly** scratch the sleepy girl?*

The Subject is also not the person or thing that does the action in a sentence. *The sleepy girl* is not the 'doer' in either:

> **The sleepy girl** *will get scratched by Lilly.*

or:

> **The sleepy girl** *was frightened of Lilly.*

but *the sleepy girl* is the Subject of these sentences: if we make them into questions, the words *will* and *was* hop in front of *the sleepy girl*.

> *Will **the sleepy girl** get scratched by Lilly?*

> *Was **the sleepy girl** frightened of Lilly?*

This shows that the meaning of the words is not relevant to the idea of Subject, whether it's what the words are being used to talk about, or what kind of role they are playing in the situation being described. The specific place of the words in the sentence—first word, second word, etc.—is also not relevant. The notion of Subject can't be reduced to meaning, or to word order.

There are other properties of words that allow you to work out what the Subject is in English. Sometimes the number of things the Subject is used to refer to affects the shape of the verb. When the Subject is used to refer to multiple things, like *Anson and Minnie*, the verb in the following sentence takes the form *fear*.

Anson and Minnie fear Lilly.

But, if we change the Subject and use it to refer to one thing, the verb changes its form to *fears*, with a final *s*.

Minnie fears Lilly.

We don't see the same change in the verb when we alter the number of individuals of non-Subjects in the sentence.

Minnie fears Dodger and Lilly.

Minnie fears Lilly.

It doesn't matter here how many people—well, cats—the words after the verb are being used to refer to. The verb doesn't change its form. In English, the form of the verb cares about the Subject.

This phenomenon, where the verb changes to track properties of the Subject, is called Agreement. We say that the verb agrees with the Subject. We see Agreement in examples like those above, and also when the verb *be* changes its form—in most dialects of English, you say *I am*, *you are*, and *she is*, and not *I are*, *she am*, and *you is*. In English, verbs agree with Subjects.

There are languages that allow Agreement with non-Subjects. My favourite of these is Kiowa, an endangered Native American language spoken mainly in Oklahoma, that I worked on with my colleague Daniel Harbour.[5] In Kiowa, a verb will show different Agreement depending on properties of not just the Subject, but also of other phrases in the sentence. Here's how you say *I gave a book to the man* in Kiowa:

náw k'yáahî̃ kút yán-áw̃

The Kiowa word for 'man' is *k'yáahî̃*, and the word for 'book' is *kút*. The *náw* at the start means 'I' or 'we'. The verb meaning is given by just the *áw̃* that appears at the end of the last word, which signifies 'give'. The rest of that verb is the syllable *yán* which signifies that the Subject is the person speaking, that the thing that's being given away is just one thing, and, that there is just a single individual receiving it. That one syllable is the part of the verb that agrees, and it agrees with everything else in the sentence, not just the Subject. We can line up Kiowa and English to make the correspondences clearer:

náw	*k'yáahî̃*	*kút*	*yán-áw̃*
I	man	book	I-it-him-give

If a bunch of people were giving someone two books, that syllable at the start of the verb would look completely different. It would have been *mé*, not *yán*. Kiowa Agreement gets pretty complex because so much of the sentence gets involved.

Kiowa shows us that Agreement with a verb is not restricted to Subjects across languages. Languages like Kiowa don't single out the Subject as something special. Languages like English do. There's an abstract property of parts of English sentences—grammatical Subject—that is central to how that language works.

This abstract property is not a symbol. It is not a link between what we see or hear and a meaning. We can't reduce it to a link between, say, the first word in a sentence and the actor in a situation. To really define what a Subject is in English, we need to look at the way that English syntax works as a whole, taking into account word order, Agreement, and many other properties of the way that English works. A Subject is a crucial part of the invisible weave of structure that makes up English sentences.

This notion of Subject is not something that is detectable in the hearable or seeable form of the sentence. It is an imperceptible property. But if a symbol is a link between a concrete form and a meaning, then the notion of Subject can't be a symbol. It neither has visible form, nor does it signify a particular meaning, yet it is crucial for explaining how English works. Language goes beyond symbols.

We've used emojis so far in this chapter as a kind of tool, as a way of thinking about how far simple symbols are from human language. I've shown you how we might augment symbols to try to capture some of the properties of actual language. In the end, even extended symbols aren't sufficient. Abstract properties, that can't be seen or heard, are an inescapable characteristic of how language works.

Is language just communication?

To answer that, we need to ask: what is communication? At first blush, we might say that communication is the exchange of information. We do use the English word 'communication' to talk about when information is exchanged, but we also use it to talk about expressing desires, feelings, orders, hopes, and all sorts of other aspects of our internal mental life. At least as we use the word in English, human languages seem to go beyond mere exchange of information.

We don't communicate only through language. We can communicate all sorts of things through fashion, painting, music, dance, and other cultural activities. We can also communicate through raised eyebrows, smiles and groans, and emojis. Some of our communication is intentional, some of it is inadvertent—think of those emails where you've cced the wrong person. Some of it is truth, some of it lies, and some of it neither. Some communication is about social status, or expectations of the moment, and much of it is unconscious. When my cat's miaowing at me for food, and I impatiently say 'Yes, yes. I'm getting it. Just hang on till I get the tin opener,' do I communicate to her? She's not a person, I'm pretty sure her miaowing isn't a human language, and I'm pretty sure she's no idea what I'm saying. In fact, she continually miaows at me in a more and more desperate fashion as I struggle to open the tin of food, so me telling her I'm opening it is definitely not being successfully communicated.

Saying that language is communication doesn't really give us much insight if we just think about what the English word 'communication' means. Can we do better than just trying to analyse the concept? Is there a way of understanding communication from the point of view of science?

There are, in fact, scientific theories of what communication is. Communication can be understood as what happens when some information gets encoded as a signal and is transmitted to something that receives it, decodes it, and thereby ends up with the message. Language doesn't need to be involved at all. A digital radio transmitter communicates information to a radio receiver by coding the sounds made in the studio as a digital signal. This is then sent zooming over the internet, or over digital radio networks, to your phone or laptop, which decodes it, and plays the music.

Human beings communicate without language too. In the Sherlock Holmes story, *the Hound of the Baskervilles*, there's a murderer living on the moors (spoiler alert!). The moors are barren and freezing and there's nothing to eat. But luckily for the murderer, his sister works in the big manor house and is married to the butler there. The butler and the sister concoct a plan to feed the murderer—the sister has a soft heart. The butler communicates to the murderer that he can come and pick up food by holding a candle by a particular window at a particular time. The murderer communicates he's got the message, by holding up his own candle, in return. All this ends up disastrously when the intrepid Dr Watson gets involved.

This butler-murderer example is particularly instructive. There's no language involved in the actual act of communication—though there probably was to set up how the communication would work—but a life-or-death message is communicated. How does this happen? It's because both the sender of the message (the butler) and the receiver (the murderer) know what the range of messages can be: it's safe to come and get food, or it's not safe. There are only two possibilities: a candle at the window conveys it's safe. No candle, it's not. Communication happens when the butler produces a signal. This is carried by light waves through the night, to the eyes then the brain of the murderer, who is able to decode it. The act of communication has an effect on what the murderer believes about the situation: his uncertainty about whether there is food to be got at the back door is reduced.

You can even lie with this incredibly simple system of communication. Imagine that someone had learned what the butler was up to, and signalled using a candle with the intention of luring the murderer to the back door to capture him. Communication would still have happened, as the murderer's uncertainty about the situation would have been reduced. Unfortunately for him, that particular act of communication would have effectively been a lie. However, it was still communication: a meaning was got across by means of a signal.

The American engineer and mathematician Claude Shannon, sometimes called the father of information theory, developed a scientific understanding of communication along these lines. At the heart of this is the idea that communication happens when the uncertainty of the receiver of the message is reduced. In our Sherlock Holmes example, the murderer has a finite set of possible messages—there are just two possible messages. Before he's seen the signal, he doesn't know whether coming to the back door to get food is going to be successful. After the signal, he at least thinks he knows. So he's received a unit of information— what Shannon called a *bit*. For Shannon, communication happens when something receives units of information and a unit of information is just something that affects your certainty about the world.[6]

Shannon's theory also allowed for what happens when the message is corrupted as it's transmitted. In our example, we could imagine that the murderer might be hallucinating, and see a candle when there was none. The message—no candle at the appointed time, so it's not safe—is not received properly because of the murderer's hallucinations. Or perhaps a gargoyle, knocked off its perch by a Dartmoor storm, blocks the line of sight from the murderer's hideaway, so he doesn't see the signal. In this case the signal is given, but not received. Shannon modelled interference like this as noise in the signal, and its effect was to lower the amount of information that the receiver gets. Less is communicated.

We certainly do use language to communicate in Shannon's sense. When I'm writing this, I'm attempting to provide information to you that reduces your uncertainty about what I think about the topics in this book. You gain information, that you can then think about, ignore, criticize, laugh at, blog about, or whatever. We can, in fact, take a well developed scientific approach to communication, like Shannon's, and say that language is used to communicate in that sense. Perhaps all of the other things we do with language which aren't strictly communication—like me talking to my cat—are offshoots of that primary fact.

In Shannon's approach, communication has happened when a signal is transmitted that changes the receiver's certainty about the world. This means that the receiver has to have a finite bunch of possible ways she or he thinks the world is, and all the signal does is shrink these down to a smaller bunch. For the murderer, there are two possible ways the world can be (safe or not safe), and the candle signal reduces these down to one.

Meaning in language doesn't work like that, though. Sentences in language create meanings where there were none before: part of the amazingness of language is its creativity, its ability to conjure up new ideas that have never been considered before. It's the engine of our imaginations. If I say to you *The giant spider knitted me a beautiful new hat*, or *A purple hippo just licked my toe*, I've not reduced your uncertainty about the world, I've created new concepts in your mind. I've created a fictional world for you.

There is another objection to thinking of communication in Shannon's sense as central to what language is. Communication is one of the things language can be used for, certainly, but just because something is used to do something, that doesn't tell us what that something is. Use isn't essence.

Alcohol—strictly speaking, ethyl alcohol—is used to, shall we say, lubricate social situations. But it is also used to disinfect wounds or medical instruments, or to ease stress or heartache. It can be used to dissolve other chemicals to make a solution (think sloe gin), to preserve foods, and it is used in thermometers because of its low freezing point. However, although alcohol is used for all of these things, the uses don't tell us what alcohol is. To know what alcohol is, we ask a chemist, who tells us that its chemical formula is CH_3CH_2OH.

Alcohol has a structure, and many uses. It occurs naturally as a side effect of fermenting sugar. Certainly, human beings and some enterprising other animals, including the chimpanzees of Guinea in West Africa, have learned to use alcohol to alter the way they feel, and, we humans have learned how to make it ourselves. But to understand why alcohol has the uses it has, we need to understand it scientifically.

For example, the reason that alcohol gets us drunk is that its chemical structure allows it to lock onto a particular kind of neural organization in our brains. When it does this, we end up with an imbalance in our neurotransmitters, and that lowers inhibitions, lowers our control over our physical actions and thought capacities, and produces the various other pleasurable and not-so-pleasurable effects of being drunk. Other aspects of alcohol's chemical structure ensure it has a low freezing point, is inimical to bacteria, and so on.

When we talk about what alcohol is (its chemical structure), and what it's used for (lots of things), these are quite distinct things. On the one hand we have the form of alcohol, which we understand by using chemistry, and on the other hand we have the functions of alcohol, what it does and what it is used for. The structure tells us *why* the alcohol does what it does. Both the structure and the use of alcohol are important in understanding what it is and how it works in human societies. Language is just the same.

Language is used to communicate à la Shannon or in some other way, without doubt, but it is used to do many other things too. Some of these might be thought of as side effects of its primary use as communication. Talking to my cat might be like this: I'm so used to using language to communicate that I still use it in circumstances where communication is impossible.

We also use language to order our thoughts, when we speak to ourselves in our heads: planning what to do next, thinking about why the things that happened took place, considering other people's feelings, motivations, and intentions.

We use language to express our own feelings and thoughts, even when no one is around to hear them. Reams of poetry, and diaries, and academic papers have been written that were never meant to be read by anyone else than their author. I have tens of notebooks full of writing that (I hope) no one else is going to see. The function of that writing is not to communicate. It is to help me to think. I'm not communicating to myself, since I can't be transferring information to me that I already have.

There are at least two broad functions for language: communication, and expressing, ordering, and even creating our thoughts. We don't really have any way of saying which is the primary use. We do, however, have ways of trying to find out what the structure of language is.

Fred Benenson's idea of translating *Moby Dick* into emojis worked as an art project but showed the fundamental limitations of emojis as a language. We can use symbols to communicate, but human languages go beyond symbols because they have abstract structure. While communication is one of the uses of language, we cannot identify what something is used for with what it is. To understand use, we need to understand structure.

I began Chapter 1 of the book by showing you that a central use of language is the ability to respond creatively to our experiences and to use language to invent new ideas and ways of thinking. With this in mind, we can ask: what is the structure of language that allows us to use it in this way?

3

A SENSE OF STRUCTURE

Massachusetts, 2014: A marijuana dealer in Middlesex County attempts to sell some drugs to an undercover police officer. This, as you might imagine, turns out not to be a good idea. Worse for the dealer, Massachusetts has a special law that applies extra penalties to drug dealers when they are plying their trade within a hundred feet of a public park or playground. Guess where the undercover police officer had set up the sting!

It's not often that grammar comes to the rescue of criminals, but in this case the drug dealer won an appeal in the Massachusetts Appeals Court.[1] His lawyer argued that the law banned him from selling within a hundred feet of a *public park or playground*, and he was actually within a hundred feet of a privately owned playground.

The actual phrase, *public park or playground*, is ambiguous between the two meanings: the word *public* might be taken just to restrict the meaning of *park*, or, of the whole phrase *park or playground*. It's likely that the legislators, when they drew up this law, didn't even notice the ambiguity, because, given what they were trying to do, it would be a pretty perverse law that allowed dealers a more lenient sentence when they were dealing drugs near a private playground, as opposed to a public one. But, irrespective of their intention, the phrase means what it means, and the dealer's lawyer must have made a good case that the perverse interpretation was, in fact, legitimate. Unfortunately for the pot-dealer, he wasn't so lucky on the other thirteen counts he was facing.

In this example, the words *public, or, park,* and *playground,* have fixed meanings. The phrase ends up being ambiguous not because of the properties of the words it is made up out of, but because of how those words are put together—its syntax. The kind of ambiguity at play in the pot-dealer situation, where the different meanings emerge from the way the words are put together, is called syntactic ambiguity. Syntactically ambiguous phrases can be represented, a bit like the chemist's formula for alcohol, by using diagrams, like this:

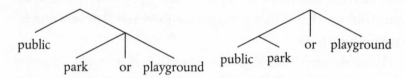

You can see that we have the same order of words in both of these structures, but the word *public* is crammed up against the word *park* in the right-hand structure. If we say that the meaning of the word *public* restricts just what it is right next to, we can explain the ambiguity of the phrase *public park or playground.*

A common sense interpretation of the law would be that it intended the structure on the right: dealers should have extra penalties if they dealt drugs in either a public park, or any playground at all. However, the Massachusetts Appeal Court decided that it was reasonable to interpret the law as intending the structure on the left: the pot-dealer would be penalized extra if he sold his wares within a hundred feet of something that was public, and was either a park or playground. Since the dealer was actually dealing near a private playground, his lawyer was able to successfully argue that the law didn't apply. The word *public* in the structure on the left applies to parks and playgrounds equally, and this therefore means that only public playgrounds are in the scope of the law. Private playgrounds are fair game for dealers!

Just as the chemical structure for alcohol explains aspects of why alcohol can be used in various ways, these syntactic structures for the phrase *public park or playground* can explain why that phrase has the property of ambiguity. In fact, when I first heard the story of the pot-dealer, I envisaged lots of new jobs for linguists, who could go through laws as they were written down, and disambiguate them once and for all, using structures like the ones above. I still think it's a good idea, but a lawyer friend of mine told me that if there were no ambiguities, lawyers would have nothing to do.

The marijuana dealer story shows us that phrases of languages can be ambiguous in their structure. None of the actual words in *public park or playground* are ambiguous in this example in the way that, say, *sty* is ambiguous (an eye inflammation or a place to keep pigs). This is why we conclude that the ambiguity comes from how the words are put together. The ambiguity is structural.

But this, if you think about it, means that there's something quite odd going on. It entails that a sentence or phrase doesn't just consist of the words that we hear or read or write down (or signs, if we are using a sign language). Beyond the words, inaudible and invisible, there's something extra that we are unconsciously sensitive to when we hear sentences. It's as though we have a sixth sense, a sense of linguistic structure, that allows us to detect the ways that words can be put together.

Otto Jespersen, a Danish grammarian and the author of *The Philosophy of Grammar*, published in 1924, writes of a child learning a language that

> without any grammatical instruction, from innumerable sentences heard and understood, he will abstract some notion of their structure which is definite enough to guide him in framing sentences of his own, though it is difficult or impossible to state what that notion is except by means of technical terms like subject, verb, etc.

Jespersen put his finger on the issue in this quote. He talks about a notion of structure that guides us when we form and understand sentences. Human beings appear to have an ability to unconsciously sense what the structure of sentences is when we hear them, though this structure is abstract. This is what allows us to judge, as we saw in Chapter 1, whether sentences are unremarkable, or somehow 'a bit odd'.

Everybody has the sense of structure that gives us the ability to make such judgments, not just people who have been schooled in the grammar of their language. I've worked with speakers of languages that have never been codified by linguists. They certainly didn't learn any grammar at school, their languages have never been written down, but they have just as firm a sense of structure as highly literate speakers of other languages. They will tell you quite firmly and consistently which sentences are part of their language, and which are not. They know when sentences are ambiguous, or have untoward meanings.

We not only sense the structure of sentences we hear or read, our sense of structure also guides us in producing sentences. Each time we turn a thought into a sentence, we give it a particular structure, a structure which connects it with meaning—hence the ambiguity of *public park or playground*—and, as we will see, with sound. Like a sculptor using their senses of touch and vision to create a sculptural form from clay, we use our sense of structure to create sentences from thought. Whoever hears or reads the sentences we construct perceives not just the sounds or letters, but also how we have created them, what invisible ties bind them together. Together with the context in which the sentence is uttered, our sense of structure allows understanding to flow.

People's sense of linguistic structure is, in some ways, not too different from their other senses. We often think of our senses as simply passively taking in information from the world. However, although we certainly perceive the world through our senses, these senses structure what we perceive.

I had a bizarre experience of this a while back at a friend's party. We were staying in a house that she had rented for the weekend, and it had a log fire. I was sitting across from the pile of wood that was to feed the fire, and I saw a face, a quite demonic face, in the woodpile. I knew consciously that it was just a collection of wood logs, red string netting, and other bits and pieces, but there was no getting away from what my brain wanted to do with it: a red demonic visage. When I got other people to sit in the same position as me, they also saw it.

That illusion arose because human brains have a propensity to interpret shapes with a face-like configuration as an actual face, even if those shapes arise from how bits of wood, netting, and so on are arranged. If you see a face, your brain also has a propensity to attribute to it all the things that usually go along with faces: intention, thoughts, emotions, etc. Hence, the spookiness of the image. My sense of vision didn't allow me to perceive sticks and wood; it created a face and that's what I saw. The ancient Greek philosopher Epikharmos of Kos wrote that 'only the mind sees and hears, all else is blind and deaf.' Although poetically expressed, this is not far off of the truth.

Similarly, when we look at an illusion, like the famous Müller-Lyer illusion, we can't help but see the lines as being different lengths, even though when we measure them they are identical. We don't consciously calculate aspects of the world, we just unthinkingly perceive them, and we have no conscious access to how that perception works.

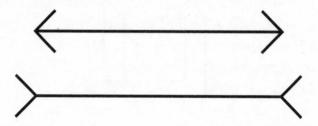

The philosopher Daniel Dennett has suggested that we have conscious access to the *results* of the processes of our minds, but we never have conscious access to the processes themselves. We don't know what the mental processes are that make us see the two lines in the Müller-Lyer illusion as different sizes unless we learn about the psychology of vision. However, we are conscious of the result of whatever our mind is doing to make them appear so.[2]

Just as we don't really have conscious access to how our sense of vision works, we don't have conscious access to how our sense of linguistic structure works. We automatically and unthinkingly know that sentences and phrases have certain properties, without knowing how we know that. The process that assigns the structure is an unconscious one.

Here's another simple example of our sense of linguistic structure that shows this. The sentence *She looked up the mountain* is ambiguous, as can be seen from two quite different ways we can continue the sentence:

She looked up the mountain (and saw tiny goats climbing its flanks).

She looked up the mountain (in her compendium of mountains).

Compare the ambiguity of this sentence to the Necker Cube illusion:

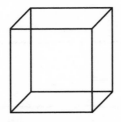

When you look at an image of a Necker Cube for a minute or so, it flips between appearing as though it is oriented down towards your left, or up towards your right. Like the linguistic example, it is ambiguous. The cube never has both orientations at once, or some kind of a mishmash between the two. It's always one or the other. This is very like our perception of the meanings of an ambiguous sentence like *She looked up the mountain*: the sentence can have one meaning, or the other, but never both at the same time, or a mixture between the two.

The Necker Cube illusion goes away if we colour one side of the cube with an opaque tint obscuring some of the lines. That signals to our sense of vision how the image should be interpreted.

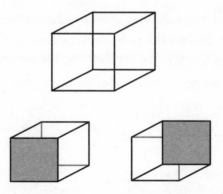

We find the same kind of effect with ambiguous sentences. If we put the word *desperately* just after *looked* in *She looked up the mountain*, only one of the meanings is possible. The meaning which involves gazing, with cricked neck, at the goats is fine:

> *She looked desperately up the mountain (and saw tiny goats climbing its flanks).*

The presence of *desperately* disambiguates the sentence so that our sense of linguistic structure only perceives one meaning, in just the same way that colouring in one side of a Necker Cube disambiguates the image. We can see this very clearly by continuing the sentence in a way that tries to force the meaning where our heroine needs to find details of the mountain in her book:

> *She looked desperately up the mountain (in her compendium of mountains).*

This is just a weird sentence. The intuitive reason is that the word *up* is more closely associated with *looked* when it means something like 'find information' than it is when it means 'perceive in an upwards direction'. But what is that 'closer association'? There's nothing visible or audible about it. Our sense of linguistic structure, working beyond the level of our consciousness, just tells us that these are the meanings that these sentences can have.

A final, striking, example of this comes from sentences like *What Anson is is silly*. Like our previous examples, this sentence has two meanings. We can see these by setting up the context in two different ways:

> *Anson is always joking around and being an idiot. If I were asked, I'd say that* **what Anson is is silly**.

> *Anson has just been appointed to the job of secretary of the new committee. But that committee doesn't even need a secretary, so* **what Anson is is silly**.

The two meanings are quite distinct. In the first, we just emphasize the fact of Anson himself being silly. In the second meaning, we're not saying that Anson is silly, but that being the secretary of the committee is silly, and that Anson has that job. Anson could, in fact, be very sensible and you could still say *What Anson is is silly*.

Not all sentences with this surface form have this ambiguity. For example, *What Anita is is proud of her garden* only has the meaning that Anita is proud of her garden. The meaning where the job that Anita has is proud of Anita's garden doesn't make sense, and so isn't present. However, when the two meanings both make sense, the ambiguity arises. For example, in *What Amelia is is important*, it could be that Amelia is important, or that whatever job she has, or role she plays, is important. The sense of structure shared by native English speakers just forces us to interpret this sentence in both ways.

But now look at the following sentence:

> *What Amelia is is important to her.*

If we take *her* and *Amelia* to pick out the same person, suddenly, one of the meanings disappears: the sentence can only mean that Amelia's job is important to her. It can't mean that Amelia is important to herself. Just like the Necker Cube or the *look desperately up* examples, the ambiguity is gone. Our sense of linguistic structure is unable to provide this sentence with both meanings. If you ask yourself why, there's no obvious answer. To provide an answer, we have to understand what the invisible structures are that underlie the ambiguity.

∞

The examples we've looked at already, where sentences are structurally ambiguous, are solid evidence for the existence of our sense of linguistic structure. There's also interesting evidence from psychological experiments that our sense of structure can't be reduced to the meaning of sentences, or how they are pronounced. The psycholinguist Kathryn Bock carried out a series of experiments starting in the mid 1980s, to test whether the abstract structures that linguists propose are subconsciously used by people as they process sentences.[3] She developed an experiment where she showed the participants pictures of people giving gifts, showing things to other people, doing things for others, and so on. Now, in English you can describe these kinds of actions in various ways. For example, you could say:

The girl is giving the book to the boy.

but you can also say:

The girl is giving the boy the book.

These sentences differ in structure, although they are close paraphrases of each other. The obvious differences are that the order of the words is not the same, and there's an extra word *to* in the first sentence.

Bock was interested in finding out whether this difference in structure was something people were sensitive to. To test this, she first explained to the participants in her experiment what the task she wanted them to do was. For example, she'd show them a picture of a girl, a boy, and a book, where the girl is giving the book to the boy. She'd say one of the sentences above as part of explaining that she wanted the participants to describe the scene. Then the participants were shown a new picture (of, say, a man throwing a stick to a dog), and asked how they'd describe what was going on.

Unknown to the participants, Bock was carefully controlling whether, in describing what she wanted them to do, she used one kind of sentence structure as opposed to another. She then noted down which kind of structure the participants themselves used to describe their picture.

In doing this experiment, Bock discovered something which no one had seen so clearly before. If Bock used the first kind of sentence to set up the task, the participants would be far more likely to use that kind of sentence to describe their picture. This was true even though the experiment used a completely different scene and completely different words. What was even more striking is that the participants in the experiment had no idea they were doing this. They made the choice subconsciously. Since the meaning of the sentences they were using was quite different from the meanings of the sentences Bock had used, and since the words themselves were different, the participants must have been accessing the abstract structure.

This experiment shows the sense of structure at play not only in understanding sentences, which is what we've seen already, but also in creating and producing them.

Bock's experiment has been done over and over again by different researchers, with many different kinds of grammatical structure, using different kinds of set-up, and doing it with speakers of different languages. The researchers always come up with the same result: people are subconsciously sensitive to the abstract structure that sentences have, and that abstract structure influences their behaviour in doing a similar task.

Bock and other researchers have also shown that you can't tie this effect down to the particular words used, for example, the word *to* in the sentences above: what matters is the structure, which isn't even pronounced. The meaning and the words can be completely different, so the abstract unpronounced structure has to be somehow there. It's something that we subconsciously, unreflectingly, impose on what we hear, just as we impose a meaning on a visual image.

We have seen good evidence from sentence ambiguities and from how people process language that we humans subconsciously attribute an abstract structure to sentences, both sentences we hear and sentences we produce. How should we conceive of this?

The way I like to think of these structures is as kinds of mental gestures. Look at one of your hands, or your face in a mirror, and make a gesture or an expression. Your hand or face has taken on a particular structure, for just a moment: your thumb is crossing your palm, or your eyebrows are raised and your lips are opened, or whatever. What structures are possible depend on the limits your anatomy puts on your hand or face. Which particular structure ends up happening depends on your intentions or reactions.

A sentence is a bit like this. You intend to say something, and your mind creates a gesture. This gesture is just a particular configuration, for a moment in time, of your brain. The structure of this gesture is limited by the rules of your language. This structure is profoundly connected to what you intend to say, as we've already seen in the pot-dealer example. It also has an impact on how the sentence is pronounced. Certain aspects of intonation depend on the structure, as do where pauses can go, as does the order of words.

The analogy with a hand gesture or facial expression isn't perfect here. There's only a limited set of gestures you can make with your hand or face, but there's a vast number of sentences that any human mind will create and understand over a lifetime. We'll see in later chapters that language works on discrete elements, but gestures are continuous. We'll also see that our mind's capacity for language is, in principle, infinitely more flexible than a hand's for gesture. The kinds of structure we see in a hand, and in language are profoundly different, but what we do with structure, whether we make a gesture or say a sentence, is quite similar.

I've been talking in terms of gestures of the mind. When I say mind, I'm just talking about the brain, but in a more abstract way. We know that the human brain must be doing something with abstract structures, as Bock's work shows that human beings are sensitive to these. Most of what linguists do is abstracted away from what the brain does. We look at languages and linguistic behaviour and see what kind of understanding we can build of that, and we are a long way from connecting particular patterns in language to particular brain signals at a detailed level. However, although studies of how the linguistic abilities of human brains work are at a very early stage, there is some good evidence from brain imaging research for the kinds of structure linguists have proposed.

In 2011, a Paris based team of researchers, Christophe Pallier, Anne-Dominique Devauchelle, and Stanislas Dehaene, used MRI scanning technology to see if there was any particular part of the brain that got more active when processing syntactic structures.[4] They did this by showing people, who had been placed in MRI machines, lists of words. Some people got lists that were just unrelated words. Others got lists where two words right next to each other could be understood as a single unit. Yet others got lists where this was the case with three words, and so on. The researchers then looked at how the participants' brains reacted. They found that particular parts of the participants' brains increased in activity in a way that was matched to the increase in grammatical structure in the lists of words the participants saw.

You might think that this isn't about grammatical structure; perhaps it's rather about meaning. To control for this, the team used an idea from Lewis Carroll's famous poem 'Jabberwocky', which starts:

> 'Twas brillig and the slithey toves
> Did gyre and gimble in the wabe;
> All mimsy were the borogroves,
> and the mome raths outgrabe.

Jabberwocky is from Carroll's 1871 novel, Through the Looking Glass, and What Alice Found There. In the strange new land that Alice finds herself in, she comes across a book containing it. After Alice reads the poem, Carroll's novel continues:

> "It seems very pretty," she said when she had finished it, "but it's rather hard to understand!" (You see she didn't like to confess, even to herself, that she couldn't make it out at all.) "Somehow it seems to fill my head with ideas—only I don't exactly know what they are! ..."

Alice's problem is that the poem is full of nonsense words: brillig, mimsy, mome raths, and so on. But Alice still gets an idea of what is going on because the poem recognizably follows the structures of English. This is because, although he used nonsense words which are not English, Carroll kept in English all the special little grammatical words and parts of words that signify the structure of the language: 'twas, and, the, all, the out in outgrabe, the -s that signifies Plural on toves, borogroves, and raths, and so on. These grammatical elements, together with the word order, are what make Carroll's poem seem like some kind of English, even though Alice has little idea about what is actually going on. In fact, if we were to change each of these little grammatical elements, I think Alice would have been completely perplexed:

> Va bright iss na slimey hindep
> rayn dance iss jiggle awns an grove
> oolya weary ro na porcupinep
> iss na little pigep machdove.

In this version, I've changed the forms of the grammatical words and word-parts, and, instead of nonsense words, I've used bits of real English. But, unless you know the original, it's now pretty unrecognizable as a set of English sentences: it reads like some English words surrounded by random sounds. That's quite different from Carroll's original, where the grammatical words are responsible for holding the meaning together. This is what allowed him to write 'Jabberwocky', and what filled Alice's head with ideas. The structure of sentences is held together in many languages by these little grammatical words and endings, and the Paris team took advantage of this.

As a second experiment, the team used a Jabberwocky-style list of words. Just like in Carroll's poem, the grammar was clear, but the meaning was impossible to work out. By doing this, they hoped to find areas of the brain that were sensitive to structure and not meaning.

And they did. A number of areas of the brain, working together, seemed to be particularly active in just the cases where there was grammatical sentence structure, even when there was no real content to the words in the sentences. One of these areas is low down at the front of the brain—it's called the Inferior Frontal Gyrus. Combining the two experiments allowed the team to show that, as the brain processes linguistic structure, particular networks of connections became active; more and more structure in the sentences presented to the participants leads to more and more neural activity in the Inferior Frontal Gyrus.

This work was complemented by research done by a team led by David Poeppel and based in New York in 2016.[5] Poeppel's team used the fact that the brain has a kind of pulse, in fact many pulses. These are called brain rhythms. The neurons in our brains work as an organized system of rhythms which tune our minds to our environment. Brain rhythms are implicated in much of what we do: walking, breathing and, it turns out, language. For example, syllables in languages have an average length, a quarter of a second. This is true no matter which language you are speaking, and it is a consequence of the rhythmic processing of language by the brain. If you try to stretch out some syllables and compress others, speech becomes much more difficult to make out.

Brain rhythms can be detected by a kind of brain scanner that measures the magnetic field around the brain. What this team did was show how the rhythms of the brain get in sync with the sounds people hear. They carefully carried out experiments, playing different kinds of sequences of sounds to people in magnetic scanners, and showing how different brain rhythms synchronize with aspects of these sounds. You can probably guess what I'm going to tell you. Certain brain rhythms synced with the abstract structure of sentences. The New York team were able to show that the syncing went beyond what could be connected to either the intonation of sentences, or the statistical frequency of aspects of sentences. This means that our brains, as we listen to sentences, get into rhythm with the abstract structure.

Slightly more scarily, the team also inserted electrodes into peoples' brains to find out the location of the bits of the brain where the abstract structure tracking takes place. They found that the parts of the brain that seemed to be the source for this tracking behaviour included, but wasn't limited to, the same Inferior Frontal Gyrus that the Paris team identified. These experiments which try to localize where in the brain abstract structure happens and how it happens are still quite limited. There's a huge amount more to learn, and we can only really take them as indications as to how the brain encodes abstract linguistic structure. But they do show quite conclusively that that's what our brains do.

There are, then, a lot of reasons to think that sentences are associated with an abstract structure. Our brains seem to be sensitive to it in particular ways, our behaviour seems to be sensitive to it as we process sentences, and the languages we speak show a great deal of evidence for this structure in the patterns they allow. We can think of abstract linguistic structure as a momentary mental gesture.

A human language, like English, goes beyond symbols and beyond a means of communication. A speaker of a language has Jespersen's 'notion of structure', rather than a list of symbols or a way to communicate. This underlies both our ability to create sentences, and our sense of linguistic structure, which we use to work out the properties of the sentences we hear.

It doesn't matter what the language is that people are speaking or signing around us. Whatever it is, we impose upon it, to the extent that we are able, the kind of structure that human language has. Our job as baby language acquirers is to work out, subconsciously, what particular variety of human language we are immersed in, but we will always be using the same basic principles.

The innate resource that we bring to bear in doing this is what the American linguist, Noam Chomsky, calls Universal Grammar. Universal Grammar is just the specialized inbuilt capacities we humans bring to bear when we are acquiring a language or languages. From Chomsky's perspective, Universal Grammar, plus our linguistic experiences, as well as our general intellectual skills, allow us to develop Jespersen's 'notion of structure' for our own language. Using this notion of structure, we are able to frame sentences of our own, and understand those of other people.[6]

Is there any reason to say that we have Universal Grammar, an innate and particularly human capacity, rather than just an ability to extract patterns and generalize them? Surely it would be simpler and more elegant if we didn't have to say anything special about human beings beyond saying that we are particularly good at learning. I provided some initial reasons to think that there is Universal Grammar already in Chapter 1. We seem to be able to judge whether something is a sentence of our language or not even when all the evidence points to us never having had the relevant experiences. But to really get evidence, we'd have to raise a child in a situation where we could completely control the language they get to hear, which would be deeply unethical. Surprisingly, however, there are natural situations which come close.

4

THE QUESTION OF PSAMMETICHUS

The idea that somehow languages are built into the human mind, just waiting to emerge, is an ancient one. The Greek historian Herodotus tells of an Egyptian Pharaoh, Psammetichus, who wanted to know who the first race on Earth was. Obviously, he wanted it to be the Egyptians, but the Phrygians were also in the running. He worked out an ingenious, though rather cruel, experiment to find out. He took two children, as babies, and gave them to a shepherd to look after, ordering the shepherd not to speak to them at all. After two years, the little boy ran out to the shepherd as he was approaching their hut, and cried out 'Bekos, bekos'. Psammetichus, on hearing this, was a little dismayed, since *bekos* was the Phrygian word for 'bread', but after that he yielded to the Phrygians claims.

Scientists are not, these days, interested in who the most ancient race is. Psammetichus's experiment assumed that, with no one speaking to the children, the language they would use would be the original language of humankind, and no linguist nowadays thinks that the sounds, words, or grammar of particular languages is built into our brain.

But Psammetichus's question is still one we can ask: what kind of language, if any, would emerge when children grow up with no language surrounding them?

This is a hugely important, and very controversial, question. It gets at the heart of the two very different views about human nature sketched in Chapter 1: are we born with specific mental powers that underpin essential aspects of our nature, or do we have a general capacity for understanding that we apply to the world surrounding us? Is what makes us human some set of abilities that come together to make *human* nature, as distinct from, say, chimpanzee or Martian nature; or is it that we have a flexible ability to tackle what the world throws at us? Perhaps more than a chimpanzee does, but perhaps less than a super-intelligent Martian. Focussing on language, do we have human language because of a particular way our mind is set up, or is it just that we're clever enough to figure out, and develop, the languages spoken around us? Does our sense of linguistic structure come from our biological setup, or is it a result of our general cleverness and the complex social interactions we experience in our day to day lives?

These two perspectives on language, and on human nature, are continually in tension with each other in the field of linguistics, and more widely.

Why is there a battle between these views? There's something very attractive about the idea that our language is learned by using very general abilities. If we don't need to say that there is a special capacity for language, we have a very simple theory. We experience the properties of the world around us, some of which involve people speaking to us, and us to them, and we learn from these. Because we humans are immersed in communication, there are a lot of these experiences, so we have a lot of material to learn from.

There's also a simple theory of learning that goes along with this view: when our brains take in experiences, the neurons in them fire up in certain ways. The more we have the same, or similar, experiences, the more the same neurons fire. This strengthens the link between experience and what the brain is doing. The brain changes over time, rewiring the connections between neurons. This approach to learning is attractive, as it looks quite similar to what neuroanatomists have seen by studying our brains. There are also computer programs that work by mimicking this kind of process. We'll meet some of these in Chapter 8. If we can learn the languages we speak in this way, then there's nothing particularly puzzling about language. It's just what Darwin said over a century ago: our more complex brain gives us a greater power to associate ideas than other animals have. The way we learn languages is the way we learn things in general.

The alternative is that there's a special structure to our brains. Something different happens when we learn a language. It's not the same as learning anything else. The structure we can see in languages when we investigate them is not learned from many experiences. Rather our brains impose this structure on those experiences. We learn the details of our local languages, but the overall pattern, the structure of language in general, is the same for each of us, part of our nature as humans. We use special, not general abilities when we learn languages. That's why we can do this, and other animals can't. That's why, everywhere in the world that there are humans, there is language. It's part of our nature, a property of our species.

How learning works in this view is also simple. Children experience people speaking around them. They are humans, so they immediately begin to impose a structure on this experience, a particularly linguistic structure. How they do this is guided by Laws of Language, which are part of how the human mind is organized. Learning our first language is a process where we subconsciously figure out which structures work best to help us both understand what we hear and express what we want to say. We match the range of possibilities allowed by the Laws of Language to what we hear around us, developing a particular sense for the structure of our own language.

One of the advantages of this view is that it suggests that not everything is possible: there are limits to structure. This means that we should expect to see similarities across unrelated languages. It also leads us to expect absences, linguistic rules that are just not humanly possible for children to learn. As we will see in Chapter 5, both of these expectations are met, so there seems to be good evidence supporting this approach. If this is right, then Darwin was wrong about language. What distinguishes us from other animals and allows us to have language is not just that we're cleverer, it's that our minds are set up differently, just as our bodies are.

If this alternative approach is on the right lines, then we will need to find out what the possible structures of human language are. That will help us develop an idea about the universal aspects of language, the nuts and bolts that allow us to build sentences to say what we mean. Each language is fiercely complex, and there are thousands upon thousands of them. This means we have to work hard to find out how each one works, how they differ, where they are the same. We need a scientific understanding of human language, not just a theory of general learning.

If we could carry out Psammetichus's experiment, we might learn something about this question. If children were raised with no language around them, would their urge to communicate lead to a system that uses the nuts and bolts of human language as we see it elsewhere? Would they have a sense of structure that we could detect?

We can't carry out Psammetichus's experiment. Aside from a lack of absolute royal power, it would be profoundly unethical. However, sometimes nature provides situations that are not far from what Psammetichus proposed, and there are some experiments that can be performed ethically and get at the core question, at least in a limited way. These experiments don't tell us what the word for 'bread' in some original language was. They do show us, though, that children who have no experience of language still show a sense of linguistic structure. Experience is important, but it is not everything.

Imagine you are a young baby, born profoundly deaf, to deaf parents who use a sign language. Sign languages have rich linguistic structure, just as spoken languages do. They are more than just systems of communication using symbols. Across the world, where there are communities of deaf people, their sign languages serve all of the social functions that spoken languages do among the hearing. More than that, sign languages share many general structural properties with spoken languages. Although British Sign Language is very different from English in its structure—and indeed very different from, say, American Sign Language—sign languages and spoken languages work in similar ways.

The sounds of words in a spoken language like English, as we'll see in the next chapter, are organized by where in the mouth they are made, by what action the tongue is making, and by various other things we do with our lungs, throat, and mouth. This organization isn't arbitrary—there's a coherence to the sound systems of particular languages. For example, you make the same movement of the tongue when you say the words *ton* and *done*. Try it. The difference is that the flow of air coming from your lungs is doing different things to make the sound *t* and *d*. This change in airflow is enough to distinguish the sounds, and hence the words. Sounds are built up out of certain actions of our vocal organs combining in certain ways.

In sign languages, there's a similar kind of organization, but rather than parts of the mouth and actions of the tongue and lungs, we have the shape and orientation of the hands, the ways the hands move, and where the sign is made. I don't mean to say that these two systems are strictly analogous in any way. It's not that the hands are doing things that substitute in for what the mouth is doing. But signs are built up out of certain actions of the body combining in certain ways.

For example, in British Sign Language, the sign for 'now' involves holding both your hands out in front of you at waist level, palm up, and moving them down and up again twice. The word for 'British' is exactly the same, but the hands are palm down, rather than palm up. Just like *ton* and *done*, most of what is going on is the same, but a simple switch in orientation of the hands—as opposed to the airflow from the lungs—changes the sign. The two kinds of languages, sign and spoken, share a similar organization, even though they use very different mediums.

Just as both spoken and sign languages use abstract patterns to organize the basic units that make up words or signs, both kinds of languages use abstract grammatical structures to organize sentences. Spoken languages group words together to make phrases and sign languages do the same with signs. Spoken languages have particular rules about what can go where in a sentence and so do sign languages. Sentences in spoken languages can be syntactically ambiguous; so can sentences in sign language. Although it is true that signs are able to resemble the concepts they are connected to more easily than sounds are, both spoken and sign languages make major use of arbitrary relationships between the concept and the form, and they do so in a structured way.

As a deaf young baby, you acquire your sign language from your parents, and other people who sign around you, in just the same way that hearing children acquire their spoken languages. Researchers have found that the same kind of language acquisition takes place, at the same kind of rate. It doesn't appear to matter if the language you are exposed to is spoken or signed, you will acquire it in a very similar fashion.

Now imagine you are a young baby, born profoundly deaf to hearing parents, who know no sign language. This is far more common than being born deaf to deaf parents, so many children are in the same boat as you. You can see your mother and the other people around you. They move their lips, but you can't hear the sounds they make. They make gestures with their hands, expressions with their faces, smile, frown, and do their best to express what they mean when they are talking to you. But none of that is human language. Their gestures are gestures, and they don't have the properties of a language. If your parents had themselves been deaf, they may have been able to use a sign language with you. But in the scenario I'm building here, your parents have no sign language. They just do their best with what they know.

You want to communicate with them as much as they want to communicate with you. And you want to express yourself. But you don't have any language around you to learn. What would you do? You'd undoubtedly use gestures to try to get across to your parents what your needs and desires are. The question I want to ask is this. Are your gestures language-like, even though you've had no exposure to any language?

Amazingly, the answer to this question is yes. It turns out that the gestures produced by children in this situation have many of the hallmarks of spoken and signed languages, hallmarks that are not present in the gestures made by those that take care of the children. Children in this situation are called homesigners, and the gestures they produce are called homesign, to distinguish it from the sign languages used by members of the deaf communities in different areas more generally. It's going too far to say that homesign is exactly like spoken and signed languages. The homesigning child is basically having to construct a whole new language on their own. However, the gestures made by such children go a long way to showing the sense of structure we ended the last chapter with.

One important point about homesigners is that they are able to create new signs for the concepts that they need, and they treat these new signs like they are, to all intents and purposes, words. In fact, this is just a normal thing that all humans do, when they don't have access to spoken language. For example, if you try to use gestures to tell a simple story to a friend, you'll immediately make up gestures to talk about things, and gestures to mimic performing actions. Now imagine doing this with made up spoken words, and no gestures. It's impossible. We humans have an amazing ability to use our bodies to communicate things about the world, and we use the same basic actions to do this—pointing, moving, miming shapes and sizes, etc.—so gesturing is something we're good at. We can communicate well through gesture, but, unless we are users of a sign language, that communication isn't a language. There are no human communities that communicate by miming to each other: we all use language, either spoken or signed. All of those languages have a syntax, particular ways that the words or signs come together to create meanings.

Homesigners, however, can't hear spoken language, and their parents don't use sign languages. This means that homesigners don't learn words. They create them from scratch, using gestures.

For example, a homesigner might put their fists together then twist them apart to signify the idea of breaking. Or they might hold their palm flat, and move it to signify giving or taking. Or flutter their fingers above their head to signify snow. Or even point to a chair their father usually sits in to signify, not the chair, but their father.

It's worth pausing for a second just to think about how natural this seems to us as humans, but at the same time, how amazing our ability to create words is. The children are making up signs for concepts that they need to communicate, and they seem to be almost unlimited in how they do this. They have a new concept they want to get over to their parents, so they create a sign for it. These signs are not just things that appear in the child's here and now. They are general. A homesigning child might have a sign for a dog, and can use that sign to express something about dogs, whether there's a dog present or not. In fact, the child can use that sign to say a dog is not present, perhaps combining it with another sign to express negation. Homesigners don't learn words like most hearing children do, they create them, from their minds and the raw sources of what they see or touch.

This is something that distinguishes humans from other animals. Other animals, such as certain kinds of monkeys, have quite rich systems of calls, where each call is connected with some aspect of the world (a snake in the grass, a leopard in the trees, an eagle in the sky), but all of these systems seem to be not just finite, but really quite limited. There's a short list of such calls. Similarly, chimpanzees have a system of innate gestures. One palm up for 'gimme', a flung out arm for 'shoo!', but again, these are extremely limited.

Children learning sign and spoken languages show that humans have a far more powerful system, creating and learning thousands upon thousands of signs or words. In Darwin's words from the quote I gave in the first chapter, we have an 'almost infinitely larger power of associating together the most diversified sounds and ideas'. In this area Darwin's idea is right, as homesigners show us. We humans can create basic symbols, links between inner meaning and outer form, even when those symbols are not around us, and we can associate these with ideas in a way that goes far beyond what other species can do.

Darwin suggested that humans differ from animals 'solely' in this capacity for associating sounds and ideas. But homesign shows us that human children go much further than that. As I've already suggested, humans have another, quite different, system for combining symbols.

One example is David, a profoundly deaf boy whose gestures have been studied by a team of linguists at the University of Chicago, led by Susan Goldin-Meadow.[1] David was one of an initial cohort of ten profoundly deaf children whose parents were hearing, and knew no sign language. Goldin-Meadow's team recruited these children from Chicago and Philadelphia. David was the most talkative of the children and would happily produce nearly 400 signs in an hour. Sometimes Goldin-Meadow's team would end up staying over three hours, video recording David and his parents interacting. They did this over a two-year period from when he was two years and ten months old, until he was four years and ten months old.

The team found that David's gesturing had many properties that were best understood as grammatical patterns, even though he wasn't exposed to any sort of language which itself had grammatical patterns.

A good example of this is that David would create two-gesture phrases, consisting of a gesture for an object, plus a pointing gesture. For example, David would point at a coin then make a sign for a coin, just as though the pointing gesture was like the word *that* and the sign for the coin was like the word *coin*. It looks like David was signing POINT COIN as a single unit of grammar in the way that the English phrase *that coin* is a single unit of grammar.

Goldin-Meadow's team was able to show that these two-gesture phrases of David's behaved like single units in his signing. This is just what happens in many spoken languages and in other sign languages. When you say:

Give me those coins!

the words *those coins* act like a single unit. So you can also say:

Give those coins to me!

but, in English, you don't interrupt the unit and say:

Give those me coins!

There's a unity to the idea of 'those coins', and a unity in how languages tend to combine the corresponding words.

David's two-gesture phrases behave like grammatical units. When Goldin-Meadow's team examined David's signing of pointing gestures plus signs for things like *coin*, or *puzzle*, the vast majority of these (93%) involved the pointing gesture and the sign being right next to each other. They were acting as a single grammatical unit, to make a single meaningful unit. There were a small number of cases where the pointing gesture and the sign were not together, but the fact that the majority were together strongly suggests that they form a grammatical cluster.

But more than this, Goldin-Meadow's team showed that the two-gesture phrases, like *those coins*, functioned in David's gesturing just like each of the words on their own. Again, this is just like spoken and signed languages. For example, in English, you can say:

Those coins are valuable.

Those are valuable.

and:

Coins are valuable.

This shows that *those coins* is a complex structure that can appear in the same place in sentences as a simple word, like *those* or *coins*. Linguists say that the distribution of all of these phrases is the same. When a complex phrase has the same distribution as a simple phrase, that suggests that you have a structure, a bit like the structures we saw with our pot-dealer example in the last chapter. It's as though there's a slot in the sentence where either a simple, or a complex unit can go. Just like we did in the last chapter, we can draw a diagram for this—again, a bit like a diagram in chemistry for a molecule of some sort:

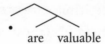

In this structure, you can either slot in the phrase *those coins* or just the word *those*, like this:

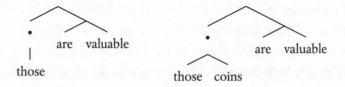

David's use of two-gesture phrases works in just this way, suggesting that David structures his gestures into the same kind of grammatical patterning seen in language in general.

To make sure that David really didn't have access to any kind of grammatical patterning, the team also analysed the gestures made by David's mother, and put the recordings of her gestures through the same kind of scrutiny that they had applied to David's gestures. It turns out that David's mother didn't make even one two-gesture phrase in all of the gestures she made over the two-year period. More recently, in 2016, Goldin-Meadow teamed up with the linguist and computer scientist Charles Yang and they ran careful statistical techniques on all of the data the researchers had collected from David and his parents. Their analysis clearly showed a huge gulf between David's gestures, which had all the hallmarks of a mental grammar with rules, and the gestures of those around him.[2] Where did David's grammatically structured two-gesture phrases come from if not from the gestures he saw around him?

Goldin-Meadow's team also showed that David's gesturing had many other hallmarks of human languages: it makes a split between nouns and verbs, it inflects words for grammatical properties, it embeds sentences inside other sentences, it has specific grammatical ways of making sentences negative, or into questions, and its word order is sensitive to who does what to whom. The team also showed that this wasn't just because David was somehow special. Other children that they studied, in the United States, in Taiwan, and in Turkey, showed the same kinds of effects. Their striking findings seem to be telling us about how resilient certain aspects of human language are. Even when there's very little language around, as in the case of homesign, still the sense of structure emerges, and the structures are those we find in languages that the children had no exposure to. This is a finding that almost seems paradoxical if we think of an individual's language as being learned from the individual's linguistic experiences. David can't be extracting patterns from what he sees around him, then using these to guide his own signing. There are no such patterns. Rather David is imposing a structure on his own signing, even though that structure is not to be found in the linguistic or gestural acts of his mother.

∞

Although Spanish is now spoken throughout Mexico, over seven million Mexicans are estimated to speak an indigenous language. One of these languages, Tzotzil, is spoken in the Chiapas region of southern Mexico, and is descended from the Mayan languages that have been spoken there for thousands of years. John Haviland, a linguist and anthropologist based at the University of California in San Diego, has been working on the culture and language of the Tzotzil people for many years. While studying the culture and language of the Tzotzil, Haviland came across a situation that would have made Psammetichus sit up and take notice.[3]

Haviland is so integrated with the community that he has effectively become a member of one of the families in a local village, Zinacantán. Mario and Rose, who head the family, already had three daughters by 1976, when Rose gave birth to Jane. Jane never spoke. She was, in the Tzotzil vernacular, *uma*, 'dumb'. Her only way of communicating with her family was through a developing homesign. Then, in 1982, Jane's brother Frank was born. Like Jane, he was *uma*. Four years later, Terry, a new sister, came along. Terry did not speak at all until she was three, and communicated with her older siblings by using signs. When Terry eventually began to speak Tzotzil, it became clear that the problem with her older siblings was deafness, and a doctor diagnosed Jane and Frank as profoundly deaf. Two years later, Will, also profoundly deaf, was born.

These four brothers and sisters used a developing homesign to communicate. The homesign initiated by Rose was enriched over time through communication between the brothers and sisters. By the time a niece, Rita, was born in 1993, the system was rich enough that Rita also used it fluently.

This family group created a sign language from scratch. None of them had ever met other deaf people. None of them had exposure to any other sign language. Haviland characterizes the siblings as 'a tiny island of signers in the midst of a sea of Tzotzil speakers, themselves engulfed by the vast ocean of Spanish speakers'. Because of his close links with the family, Haviland has known all of these siblings since they were born and, in 2008, he began to study the system of communication that they invented.

The Zinacantán signers give us an insight into how quickly the homesign created by Jane and Frank evolved. Jane and Frank grew up having to develop their homesign, which fed into Terry and Will's signing.

The Zinacantán sign system quickly evolved to have differences between nouns and verbs, a fairly systematic word order, and other hallmarks of language. Haviland's work has concentrated on how the grammatical difference between nouns and verbs appeared in the signing of this family.

What is the difference between a noun and a verb? In a language like English, the distinction is connected to, but not the same as, that between things and actions. The old idea that you can define nouns as words that name persons, places, and things has some truth to it, though it's not quite right. For example, nouns can easily refer to actions. In fact the word 'action' is a noun!

Similarly, the idea that a verb can be defined as a 'doing' word, that is as a word that describes an action or an experience, isn't quite right either. If I say *Anita has a cat* the verb is *have*, but this doesn't describe an action, or an experience. Sometimes people say that a verb can also describe a circumstance or situation, so does the verb in this sentence describe a circumstance? It seems more accurate to say that the whole sentence describes that circumstance. The verb, on its own, doesn't really seem to say very much.

In fact, much like the idea of grammatical Subject, we can't really define noun or verb without a fuller picture of the grammar of a language. However, the distinction between nouns and verbs appears to be universal: the grammar of all languages is sensitive to this distinction, though exactly how it is sensitive might vary from language to language.

The word *emotion* is a noun that we use to talk about feelings. We can use this noun in a simple sentence like:

Emotions are important to an actor.

Here we see two ways that the grammar of English is sensitive to the noun verb distinction. First, the word *emotion* can add an -*s* and that affects the number of emotions we are talking about. Secondly, this word can appear as a grammatical Subject. Subjects affect the verb, and in this case the verb changes between *is* and *are* depending on whether there's one or more than one emotion being talked about: *Emotion is important to an actor.* This is the Agreement phenomenon we met in Chapter 2.

We can't, however, say the following:

I emotioned all day.

That is, *emotion* works as a noun, but not as a verb. This is why you cannot add -*ed* to make it past Tense.

Conversely, *emote* is a verb. It's comical to say:

My emote is fragile today.

But fine to say:

I emoted all day.

Different words have different grammatical behaviour. *Emotion* is a noun, and *emote* is a verb.

English is pretty flexible in the grammar of nouns and verbs—think of a word like *table*, which works well in both *The table is broken* and *I tabled a motion*. But English grammar clearly cares about the difference. Some words are treated by the rules of English as being in one grammatical class, nouns, and other words are verbs.

This difference between nouns and verbs is about how the grammar of English works. It's not about a difference in meaning between things and actions. The distinction between nouns and verbs appears in the grammar of language after language. We'll see how this distinction appeared in the Zinacantán homesign, but first let's have a quick look at another language, Nuuchanulth, where at first glance you might think there's no noun verb distinction.

Nuuchanulth is an indigenous language spoken on the west coast of Canada.[4] In this language, almost all words work like English *table*. Here's how you say 'The man is working' in Nuuchanulth:

mamuuk-maa quuʔas ʔi

The word *mamuuk* means 'work' and the ending *maa* signifies that there's just one person working. *quuʔas* means 'man'; and *ʔi* is the equivalent of English 'the'. Nuuchanulth grammar puts verbs at the start of a simple sentence expressing a statement, unlike English. If we align the words, we get:

mamuuk-maa	*quuʔas*	*ʔi*
work-s	man	the

But Nuuchanulth speakers can also say this:

quuʔas-maa mamuuk ʔi

Now we've swapped around the 'verb' and the 'noun'. We put the word *quuʔas*, which means 'man', in the grammatical spot for verbs. That word then takes on the verbal ending *-maa*. We have also put the word meaning 'work' where the noun usually goes, with the Nuuchanulth word for *the* after it. This looks like it literally means *The work mans*, but what it actually ends up meaning is the same as the English sentence *The one who works is a man*.

This flexibility pervades the grammar of Nuuchanulth, which is one of the best examples we have of a language that seems to care little about the noun verb distinction. But even in Nuuchanulth there are still differences between some nouns and verbs. For example, a proper noun, a name like *Jack*, can't appear in the initial position in a sentence. Even in a language as flexible as Nuuchanulth, the grammatical rules treat some words as only nouns. In fact, linguists haven't found a language yet where there's a completely convincing lack of a noun verb distinction. It seems like a very natural way for a language to work.

We don't define noun versus verb in terms of meaning. We define noun and verb in terms of the grammar of a language. As far as we currently know, every language makes a grammatical distinction that looks like a noun verb distinction. That difference might be a good candidate for the kind of structure that a human child's mind imposes on its linguistic experience.

One of the important differences between nouns and verbs that we see across languages is the use of a special class of elements called classifiers. Classifiers pick out some important property of a thing: its size, its shape, what it's made out of, its location, what you use it for. Classifiers are found in many languages. For example, the language spoken by the Tlicho people of the Northwest Territories of Canada, puts classifiers on verbs to signify a property of a noun that occurs with the verb. Look at these two sentences in the Tlicho language:

let'e niyeh-ʔa

let'e niyeh-tshi

The word *let'e* is 'bread', while *niyeh* means 'to pick up'. But the first sentence means 'to pick up a loaf of bread', while the second means 'to pick up a slice of bread'.

This is because the endings of the verb classify the noun in the sentence. *ʔa* is used when it is round, while *tshi* is used when it is flat and flexible. It's as though the Tlicho speakers are saying something like:

I picked up something round which was bread.

versus:

I picked up something flat and flexible which was bread.

The first of these is, naturally, used for loaves of bread, and the second for slices.

In other languages, the classifier appears right next to the noun it classifies. Yidiny, an aboriginal language spoken by the Yidinji people of North East Queensland in Australia, is a good example of this. To express *the girl saw the wallaby* in Yidiny, the most natural thing to say is:

bamaal yaburunggu minya ganguul wawaal.

Here *bamaal* is the word for 'person', and *yaburunggu* is the word for 'girl'. *Ganguul* means 'animal', while *wawaal* is the word for 'wallaby'. Literally, this sentence translates as 'The person girl saw the animal wallaby.' Yidiny speakers use their classifiers constantly.

Classifiers are a hallmark of the noun verb distinction in many languages. They might appear next to verbs (as in Tlicho) or next to nouns (like in Yidiny) but their job is to tell us something about a noun: the size, shape, use or location of what the noun is being used to talk about.

Haviland discovered that classifiers were invented by the Zinacantán signers as a means of distinguishing verbs and nouns in their homesign. Nouns in the Zinacantán homesign are accompanied by classifiers. In fact, classifiers are used to make sure that what looks like an action when it is signed is grammatically understood as a noun.[5]

A striking example is the word for 'chicken'. Haviland showed the youngest Zinacantán homesigner, Will, a picture of two chickens, and asked him to sign this to his siblings.

What Will does when he makes this sign is first hold his two hands, loosely cupped, towards each other, roughly in the size that a chicken would be, as you can see in Figure 1. This is the classifier and it's a classifier for size and shape. Then Will makes the following action (Figure 2): he holds his two hands together as though loosely grasping something narrow and cylindrical, then he sharply moves them apart, with a jerking gesture. Why? Because that's how you kill a chicken in the village. You grasp its neck, then snap it. Then Will holds up two fingers to signify how many chickens there are.[6]

Figure 1 Will signing the classifier

Figure 2 Will signing *chicken*

There's a grammar to this: the first sign is a classifier, signifying the size of the kind of thing you're talking about. The next is an action that would be typical for that thing. The action is conventionalized. There are lots of other ways that the idea of chicken could be signed—I'd have flapped my arms by my side, I think. The final sign specifies how many things are being talked about. If this were a spoken language, it could be pronounced something like the following:

medium-size-thing chicken two

The sign for chicken here is not a pantomime: it's part of the basic vocabulary of the language. When Haviland asked Will to sign two baby chicks, Will made exactly the same series of signs, but now the classifier he made involved him holding his thumb just apart from his fingers, to signify 'small' or 'tiny'. The rest of the signing was the same, even though chicks are never killed by snapping their necks. The sign for chicken is a convention, and the ordering of the signs is part of a grammar.

Another beautiful example of the way that classifiers are used to make the noun verb distinction in the Zinacantán homesign is the sign for 'hammer'. It's generally true that with objects we use as instruments to do things, the difference between action and object is very difficult to display using gesture. Try to sign, to someone, the difference between 'I see hammers' and 'I see hammering happening'.

The way the brothers and sisters in the Zinacantán family resolved this issue is that, when they intend the noun meaning, they use a classifier signifying the size of something held then one hand hits the other. To signify the verb, there is no classifier, and the hammering action is performed not by one hand hitting the other, but more like an imitation of the actual action: two hands grasp an imaginary hammer and bring it down on something.

The crucial thing here is that the noun and the verb are distinguished grammatically: there's a grammatical rule that requires a classifier to be used with a noun. A verb can be distinguished from a noun by the presence or absence of the classifier.

But now comes Psammetichus's question. If this is a grammatical convention, where does that convention come from? It's not in the gestures of the hearing speakers around the children as they grow up. It's unlikely to have come from the one hearing sibling who also signed (Terry), as classifiers are present in both Jane and Frank's homesign, and they were born well before Terry. Classifiers, and the special syntax associated with them, appear to have grown out of nothing.

Classifiers appear in many spoken languages. It's striking how they emerge spontaneously and naturally in the Zinacantán homesign with a syntax that is very similar to the syntax of classifiers more generally.

Although these studies of homesign are not exactly Psammetichus's experiment, they come close. They show that homesign, a system used by people who have had no linguistic input at all, has certain properties of other spoken or signed languages. We saw that David's homesign clusters signs together into units, just as spoken or sign languages do. We also saw that homesign recreates the distinction between nouns and verbs, which appears to be universal to human language. Researchers working on homesign have found many other hallmarks of spoken and signed languages, even though the homesigners have no experience that would explain this. These findings are difficult to understand if acquiring a language just involves extracting patterns already present in the language surrounding us. In the case of homesign, there are no such linguistic acts. There is communication by the parents and other caregivers, using gestures, but these gestures have no special linguistic structure—they do not cluster together into grammatical units. However, homesigners' own linguistic acts do show the kind of structure we find in spoken and signed languages. The homesigners are creating their language from their own internal resources.

Homesign is quite different from how language is usually acquired by children. It's this difference—the lack of any language for the homesigner to use as a model for their own signing—that has allowed Goldin-Meadow and her fellow researchers to probe what happens to language when the child has such drastically reduced information. So what would happen if you took a lot of homesigners, who have all developed different homesigns in their own families, and you brought them together so their homesigns interacted?

This is exactly what happened in the by now famous case of Nicaraguan Sign Language, usually called, in Nicaragua, *Idioma de Señas de Nicaragua* or ISN. Deaf schools were not established in Nicaragua until the late 1970s. One particular school in the capital, Managua, had been established as a private school, that, with the success of the Sandinista revolution, was made public as part of the new government's policy of education for all. This school brought together children and teenagers who had been previously living in their own homes with their families. These young people had been communicating using different, often idiosyncratic, homesigns with their families. These children came from families where the parents were hearing, and hence didn't use sign languages, just like the children in Goldin-Meadow's and Haviland's studies. The Managua school advocated education for the deaf through spoken language (Spanish), attempting to help their students to lip-read. What happened instead was that the students interacted with each other, using the system of communication that was most natural to them, the homesigns that they had developed with their families. Many of the signs bore a resemblance to the things or actions they referred to. This meant that, even though different children used different signs, they could make good guesses about what each other meant. From these interactions, the children began to converge on a linguistic system that was usable by everyone, a common language.

At first, the sign language used by the children was rudimentary, compared to fully fledged sign languages, but even these early stages of the emerging languages began to show immediate signs of quite sophisticated structure. Following on from work by Judy Kegl, who was one of the first to document the emergence of a sign language in the school, a team of linguists, headed by Ann Senghas, explored how the signing of the students worked.[7] The researchers showed the same video to different students in the school, then asked them to tell the story afterwards, using their signs. The results were quite remarkable, even in this first generation of signers.

In English, if we want to say something like *the girl tapped the boy*, the order of the phrases in the sentence gives us an unambiguous sign of who is doing what to who. If I say *the girl tapped the boy*, you know that it's the girl who is doing the action, and the boy that is being acted on. This sentence just can't mean that the boy tapped the girl. English is quite strict about this. If I say *the cup tapped the boy*, even though cups can't usually tap things, the sentence still means that what happened was a cup somehow tapped a boy— perhaps we're in a fairy story or a dream.

What Senghas and her co-researchers found was that the signs for an action in the first generation of signers always came at the end of the sequence of signs. So to say *the girl walked*, just as in English, the signer would make the sign for a girl, then the sign for walking. However, if we have something like *the boy tapped the cup*, the signs for cup and boy both have to come before the sign for the action of tapping. Using capital letters for signs, we'd translate *the boy tapped the cup* as:

BOY CUP TAP

The most interesting case is what happens when two people are involved in the action (rather than a person and an inanimate object). What the signers did then was to split the verb into two parts, one signifying an action, and the other the result of that action. For *the girl tapped the boy*, the signers would make signs in the following order:

GIRL TAP BOY TAPPED

Here the signer would make a tapping action (which I've just written here as TAP), and a being tapped action (written as TAPPED). So when two living people are involved in the situation that is being described, we end up with two aspects of the verb being signed, one of these (the first) signifies the person doing the action, and the other the person who the action is done to.

This structure is a little like what we see in English when we use sentences like *I wiped the table clean*. I'm doing some wiping of the table and the result is that the table ends up clean. The boy in the signed example is affected by the girl tapping him, and ends up being someone who is tapped.

These types of structure have been intensively studied by linguists, and some spoken languages make much use of them. For example, in Yoruba, a West African language, we would say the sentence *Femi pushed Akin* (where Femi and Akin are names of people) as:

Fémi ti Akín subú

Here the word *ti* means 'push' and *subú* means 'fall'. So literally we have:

Femi push Akin fall

The first generations of ISN signers used a structure that looks very much like this. This is a common linguistic construction which is found not only in Niger-Congo languages like Yoruba, but also in Chinese, various native South American languages, and languages of Indonesia. But we don't find it in Spanish, the main spoken language of Nicaragua. Again the question arises: where did this structure in ISN come from?

As new generations of younger children joined the school, they interacted with the older students and, with no instruction, began to develop the language further. These new signers changed the Yoruba-like structure where the two parts of the action are separated by a noun. Instead, they began to use the space around them to signal who is involved in the situation described by the sentence. They would first set up the individuals involved in the situation by signing nouns for these before signing the verb. For our example, *the girl tapped the boy*, a signer would make the sign for girl to the right of their body, and the sign for boy to the left. The space in front of the signer, where the signs were made, was used to mark who was doing the action, and who was on the receiving end. The Subject of the sentence is signed to the right, and the other noun to the left:

GIRL(right) BOY(left) TAP-TAPPED

In fact, the second and third generation learners went even further than this. They used a pointing action to mark where the two nouns were signed (to the right and to the left) then they incorporated those pointing actions into the verb. So the example looks more like this one, where R and L are pointing actions to the right and left of the signer:

GIRL(right) BOY(left) TAP-R-L-TAPPED

As you can see, quite a complicated and rich structure emerged in the grammar of ISN. In fact, the final structure bears a striking resemblance to the phenomenon of grammatical Agreement. In English, a Subject that refers to one (Singular) versus more than one (Plural) thing changes the shape of the verb (*The girl sings* vs *The girls sing*). In other languages, other nouns aside from the Subject can also affect the ways that the verb looks. We've seen a case like this already in the Kiowa example from Chapter 2. ISN, by the third generation of signers, began to work like Kiowa. In both languages the nouns that tell you who or what is involved in the sentence come first, then these nouns are linked to the verb by putting a reduced set of information about the nouns on the verb. Here's the Kiowa sentence we saw from Chapter 2 again:

náw	k'yáahî î	kút	yán-áw̃
I	man	book	I-him-it-gave

This means *I gave the book to the man*: *náw* is the word for 'I', *k'yáahî î* means 'man', and *kút* is 'book'. These nouns are stated first, then the syllable *yán* appears attached to the verb *áw̃*, which means 'give'. That syllable encodes that the Subject of the sentence is

I, and that the other two nouns are both Singular. This is very similar to what ISN does. The relevant nouns are placed in space, one to the left and one to the right, but both are signed before the verb is signed. The verb itself has, attached to it, a sign that connects to the spatial position of those two nouns. ISN uses the directions right and left to add in extra grammatical information, and spoken languages can't, by virtue of being spoken, do this. But otherwise, the similarities are striking.

No one made up these rich grammatical complexities in ISN then enforced them. No one told the ISN speakers about the structure of Kiowa, or Yoruba. What happened was that the rules emerged naturally from the interactions of the signers with each other. These rules also bear an uncanny resemblance to abstract properties of certain spoken languages on entirely different continents.

New generations of signers have come to the school over the years, and ISN has become more and more established, to the point where now, only a few decades after the very first speakers attended the school, there are dictionaries and grammars of the language. These are still in development, because the range and complexity of the language is already so vast that it goes beyond the resources of teams of linguists to figure out all of the intricacies.

There's an interesting extra twist to this story of the development of richer grammatical systems in ISN: it's the youngest children—those who are five or six years old—that play the strongest role in restructuring the language. The various groups of researchers who have worked on ISN have watched it being born and developing over the years into a full functional and richly structured language. These researchers report that the youngest signers enrich the language with new devices, but these are rarely taken up by the older signers. The oldest signers are still using a version of the pared down system they started with, while the youngest signers have access to the kinds of more complex structures just described.

The Nicaraguan situation is fascinating. If the findings on homesign I discussed above are right, then the situation worked like this: various homesigners brought their own systems together in the late 1970s, systems already partly moulded by an inborn ability to create certain kinds of linguistic structure from raw experience. These different, idiosyncratic systems were then put in a situation where they interacted. The children just wanted to connect with each other, and the means they had to connect was their homesign systems. The rich social interactions that the children engaged in created free flowing language around them. Their capacity to create grammatical structures from the richer and richer seams of raw language they were surrounded by, together with their need to communicate, to be understood, and to express what they wanted to say, all interacted to create, in an incredibly short period of time—only a decade or so—a fully functioning human language.

The ISN situation also gives us an insight into what comes from our biological nature as human beings, and what comes from social interaction. Each homesigner comes up with a solution to their need to express their thoughts through their signing. That solution is their own, special, system of signs. These systems, we have seen, include patterns of syntax that we see in other human languages. In fact, they have the hallmarks of human language in general. Homesigners' solutions to their need to communicate have rich abstract structure, for example. That's not to say that all homesign systems are exactly the same. They are not. Each child solves the problem in a slightly different way. But they come to that solution using an ability to create linguistic structure that is common to all of us.

But we also saw rapid change in the systems used by the children when they came together in the school in Managua. In communicating with each other, they moulded and shaped their own individual systems, so that they could do what they wanted with their language. This meant that individuals were using slightly different systems, perhaps borrowing ways that other people expressed things, perhaps making mistakes that took root. No doubt some signers were sources of linguistic innovation, perhaps because they were particularly expressive, or because they were the leaders of their groups of friends. Each individual was bringing something creative to the table. This variation in signing is at the heart of the changes that took place. They provided the material for the new language to develop.

Looking at the birth of this new language surely tells us something about how the first human languages developed. The human capacity to create linguistic structure provided a space in which children could find a solution to their need to express themselves. As these children engaged with the linguistic, social, and natural world around them, their language developed in constrained ways, fed by their interactions with others and their own capacity to innovate. It became, at some point, probably very quickly, a fully developed language that provided new generations with a rich linguistic environment in which they could develop their own sense of linguistic structure.

Homesign and ISN strongly suggest that any adult who fluently speaks or signs a language has not learned that language by just abstracting patterns from the linguistic acts they have experienced. Rather they have subconsciously imposed particular patterns on what they have heard or seen, and over time have developed their sense of structure. This sense of structure is what allows language to be used for communication, for planning and thinking, and for everything that these entail. It's also the structure that gives us syntactic ambiguity—remember the pot-dealer example. It explains why groups of signs act together in David's signing, even though they don't in his mother's, and why abstract grammatical notions, like noun versus verb, are at play in the different homesign systems of children not exposed to spoken or sign languages.

∞

I began this chapter with Psammetichus's experiment. The king was interested in finding out who the most ancient people were, but we're more interested in what the human mind brings to the acquisition and use of language. One way to catch a glimpse of this is by looking at people who acquire language in a situation where they have no language around them. Human beings want to connect with other people around them, and every human society uses a language to do this. There's something deep within us that makes us want to use language in this way—humans are very social animals. Homesigners have this need, but don't have the social resources to draw on as they have no access to the spoken language of their families. Instead they use the raw non-linguistic material that surrounds them: gestures their parents make, the shapes and actions of the things they see. From these, they create a linguistic system, albeit a restricted one. The most amazing thing is that, even though their parents' gestures are unstructured, the homesigners' gestures have a number of the most important hallmarks of human languages. Most crucially for the argument I'm making here, homesign displays abstract syntactic structure. This structure has to come from somewhere, and it doesn't come from the language that surrounds the home-signers, which they can't hear, or from their parents' gestures, which have no such structure. I've suggested, following Goldin-Meadow, that it comes from within the child.

I've argued in this chapter that there's an ability to create a system of linguistic structure, universal to human beings, that imposes a particular kind of pattern on our experience of language. As mentioned in the last chapter, this is what Chomsky calls Universal Grammar. Universal Grammar allows us to lock onto particular aspects of our experience, categorizing these as linguistic, and it allows us to create a system that gives those linguistic categories structure. The evidence from the homesign research suggests that this property of our minds isn't just a general capacity to pick up on patterns in the events around us. For the homesigners, there were no such patterns. The children imposed structure, to create something linguistic, which is why the term Universal Grammar—rather than Universal Pattern Extractor, or something like that—is a good one.

The term Universal Grammar refers to an aspect of the ability we all have to turn bits of unstructured experiences into a structured system of language. Saying that there is Universal Grammar isn't the same as saying that all languages will look the same, or that we have the grammar of a particular language in our brains at birth. Our modern versions of Psammetichus's experiment don't show anything like that. Rather they show that there are certain abstract principles that engage with our linguistic experiences as we grow up. There are certainly other aspects of our mind and of the world that come into play as we acquire language. Universal Grammar is necessary, but it doesn't work alone.

Universal Grammar is also involved in categorizing aspects of our experiences as linguistic.

In 2014, researchers in Yale and New York reported that they had played sounds to very young babies (one to four months old) and measured their brain responses using an MRI scanner.[8] The researchers played speech sounds, as well as non-speech sounds like laughter or coughs, to these babies, and the scans showed that the babies processed these different kinds of sounds in different parts of their brains—essentially, towards the left-hand side of the brain—already within the first month of their lives. As the children develop, parts of the brain become more and more specialized to human speech. But since all the children had this same bias towards the left-hand side of their brains, this strongly suggests that there is something inbuilt going on. Aspects of the babies' brains are keyed to interpret some experiences as linguistic incredibly early on.

I mentioned above that humans and other animals have general abilities to extract patterns from their experiences, including experiences with sequences of sounds. We will meet some of these in more detail in Chapter 6, where we will see that humans and other animals, in fact, are quite good at this. It's possible, and perhaps probable, that such abilities also play a role when we acquire language. What makes them different from Universal Grammar is that they are not specialized to language or to our species: humans and other animals use these same abilities to detect statistical patterns in all sorts of experiences that we and they have.

There's also one other factor that will almost certainly come into play. Humans are part of the natural world, and the natural world is governed by certain laws. Living things especially, which strive against the generally increasing disorder of the universe, do so through the creation of structures of various sorts, structures which are governed by natural law as well as by evolutionary processes. My favourite example of this is the Romanesco cauliflower in Figure 3.

It's possible to write a mathematical equation which will capture how the whorls of this vegetable grow.[9] You can see that the

Figure 3 A Romanesco cauliflower

cauliflower is composed of structures that mimic the shape of the whole, and these structures themselves are composed of smaller structures, again similar in shape. The laws of nature and growth interact with matter to give you something you can roast in the oven with salt and paprika. I'll argue in Chapter 9 that the syntax of human language has this same property. The clusterings of structure we have already seen in language have the property of self similarity. It is this that allows language to be used in such a limitless fashion.

If there were no universal structuring principles for language, no Universal Grammar, we'd expect that languages could vary hugely. As American linguist Martin Joos put it in 1957, summarizing the prevalent view of the time, languages will, at least potentially, 'differ from each other without limit and in unpredictable ways', because of the diversity of the experience of their speakers.[10] More than 50 years later, in 2009, Nick Evans and Steven Levinson made the same claim: there are no universals in human language.[11] In their words, 'the more we discover about languages, the more diversity we find.'

I disagree. In fact, the more we discover about languages, the more we see that the basic principles by which languages work are the same in language after language. There are certain building blocks that languages use, and certain ways that they put them together.

There are also core parts of human thought that are never used as the building blocks for the grammar of languages: they are not part of the nuts and bolts. Linguists have, for example, never found a language where the concept of the third word in a sentence is grammatically important. But we know that humans have an ability, which psychologists call subitization, to immediately perceive that there are exactly three objects in a sequence in front of them, without counting. Why is this ability not put to use in any language?

There are all sorts of pretty reasonable ways that you could create language that we never find in actual human languages. I could easily design a language that would be completely alien, and unlearnable by humans. In fact, thinking about inventing languages can give us good insight into how natural languages work.

5

IMPOSSIBLE PATTERNS

E arly in 2015, I was sitting in my office, reading a student's thesis, and the phone rang. It was a TV company, with an odd request. 'Could you', the producer on the line asked, 'make up a language or two for us?'

The TV series was a version of *Beowulf*, the Old English epic, which ITV was turning into a major fantasy series. They needed, as is almost de rigeur these days, languages for their otherworldly creatures, and for some human tribes that the heroes and heroines were to interact with.

It's worth asking why they wanted a made-up language rather than just giving their actors some random sounds to say. And what is a language when it's made up by a person, rather than just being part of our everyday surroundings? And how on earth would you go about inventing a language?

The reason that the TV company wanted an invented language, rather than just using random sounds, is authenticity. We humans live, continually, in a sea of language, and we are very attuned to it. When we hear an utterance of random made-up sounds, at least over a period of time, we can tell whether it has the coherence of a language, whether its sounds, words, and grammar are language-like or not.

Believability is a crucial aspect of a show's commercial success, and using workable functioning languages, subtitled in TV shows and films, has become, in recent years, part of achieving believability. Producers of science fiction and fantasy shows have learned this over the years, from *Star Trek*'s Klingon, through the Elvish and Orcish languages in Tolkien's Middle-Earth, to the more recent use of Dothraki in the *Game of Thrones* TV series. If a series takes off, then there are sci-fi conventions, and hordes of earnest fans, some of whom are fascinated by every aspect of the show, including the languages. For some fans, it's important that the languages used in the show are as real and authentic as possible. That authenticity is part of creating a coherent, believable world. Constructed Languages, or Conlangs, have become an integral part of TV, online gaming, role-playing gaming, and movie culture.[1]

Most of the languages of the world are not made up. It's a deep mystery, still, how the capacity for language evolved in the human species, but however that happened, languages have since changed across time, as communities of humans have spread across the world, intermarried with each other, invaded each other, and communicated with each other. This has led to thousands of different languages, some as different as Chinese and French, some as close as Dutch and German.

But languages do not vary randomly. They have a design, a structure, a pattern, in common. They are made up of particular building blocks, and there are particular ways they put those building blocks together: the rules of the language. Languages have organized patterns of sound, of words, of sentences, and though they vary in wild and wonderful ways, it is possible to discern what the rules that construct these patterns are.

So how do you go about making up a language? You first need to understand what the building blocks of human languages are, and what patterns they fall into. Then, depending on how weird you want the language you are making up to be, you use what you know about these patterns to create rules for the new language. If you want something that sounds very much like a human language, you can stick pretty closely to the basic building blocks we know from natural human languages. If you want something less familiar, you can change these basic building blocks, or the rules which languages use to arrange them, to create something that, while still a recognizable language, gives you the sense of strangeness you want. If you know the rules, you know how to break them.

For example, the linguist Francis Nolan, who made up the snake language, Parseltongue, for the Harry Potter films, wanted something that sounded alien, far from human language, and very snakey. To do that, he restricted the range of sounds, allow-ing only those that don't need you to vibrate your vocal cords in your throat—presumably lacking in snakes! When you make particular sounds, or if you whisper, your vocal cords switch off. Only using sounds where your vocal cords are turned off creates a sibilant, whispery feeling. To get a sense of the difference, put two fingers on your throat, at the front, right in the middle then read out a sentence from this paragraph. Now whisper it. You should feel a difference in the vibrations in your throat. That's your vocal cords switching on and off. You can see this even more clearly by focussing on a particular pair of sounds. Make a long ssss sound then turn it into a zzz sound, with your fingers on your throat. You should feel the vibrations turning on and off, and you can switch back and forward, sssssszzzzzzzsssssszzzzzz. Or you could try a bit of poetry. Here's Edwin Morgan's 'The Siesta of a Hungarian Snake':

s sz sz SZ sz SZ sz ZS zs Zs zs zs z

Francis Nolan created Parseltongue to sound like no normal human language by selecting only whispery, sibilant, sounds. But he still gave the language a particular organization of these sounds, together with particular forms of words, and even a special sentence structure. The sounds are organized in a pattern, as are the words and sentences. If Nolan had not done this, Parseltongue just wouldn't have sounded like a language, and it wouldn't have been convincing.

One of the languages I made up for *Beowulf* is spoken by creatures with no lips—just very big teeth. So I had to cut out all the sounds that use lips (p, b, m, w, f, v), which reduced, quite drastically, what I could use. But the producers also wanted something that English speaking actors could pronounce, so I couldn't introduce too many new sounds. This was a challenge. I had to use English consonants but with a whole bunch of them cut out. At the same time, the language had to sound otherwordly and very different from English, as it was to be used by a group of non-human creatures called Warigs. The brief was to make it sound harsh, violent, and, well, monstery.

To design the sound patterns of the language of the Warigs, I took the consonant sounds of English that are made at the back of the throat and a few at the front of the mouth where the tongue is close to the teeth. I used a sound made with just the vocal cords, the glottal stop sound used by many people in Britain to pronounce the middle 'tt' of 'butter', or the last 't' of 'cat'. Another word for the vocal cords is the glottis, and since to make this sound you stop the air coming through your mouth, we get the term glottal stop. I also used the k sound, which is made by blocking the back of the mouth with the back of the tongue, and the t sound made by hitting the area just behind the teeth with the front of the tongue. These two sounds also block the air coming through the mouth, and so they are also called stops. I added to these hissing sounds, which linguists call fricatives, made at the same points in the mouth. The sound *s*, made by the front of the tongue, the sound *h*, made right at the back by the vocal cords narrowing, and the sound *kh*, which is like a sort of turbulent *k*, and is the sound at the end of Scottish pronunciations of words like 'loch'. Putting all this together, we get a systematic pattern of sounds that we can put in a table:

	Glottis	Back	Teeth
Stop	ʔ	k	t
Fricative	h	kh	s

The symbol ʔ is the letter linguists use for the glottal stop.

To these I added sounds that are made in the same places and in the same ways but where you vibrate your vocal cords at the same time. This is known in the trade as 'voicing' the sounds. So when you voice *t*, you get *d*. Voicing *s*, gives you *z*, and voicing *k* gives you *g*. My original version of the language also had a voiced version of *kh*, which I wrote as *gh*, which is a sound that doesn't really appear in English, but is found in other languages like Scottish Gaelic, or Greek. The producers thought that that was a bit too far from English sounds for the actors, so I had to get rid of it, leaving a gap in the system. You can't voice the glottal stop (since making it blocks the vocal cords completely, but voicing requires there to be air going through the vocal cords), and voiced *h*, although found in some languages such as Arabic, is again a bit far from English. So that extends our table a bit. I've paired up the voiced sounds with their unvoiced counterparts:

	Glottis	Back	Teeth
Stop	ʔ,-	k, g	t, d
Fricative	h, -	kh, -	s, z

This way of designing the language is a coherent patterning of sounds plausibly made by creatures with no lips and big teeth. For some variety, I added in the sound *sh*, and its voiced counterpart *zh*, the nasal sound *n*, and the sounds *r* and *l*, all of which only appear voiced in English.

As well as selecting a coherent set of sounds and creating restrictions on how they could pattern together, I also set up a system of rules for combining the sounds into syllables. These rules only allow certain combinations of sounds to make up a syllable in this language. Languages in general like to restrict the kinds of syllables they use. You can see this in English, where blap is a possible syllable, but dlap, just isn't. I set up a similar set of restrictions in the language of the Warigs. By selecting these coherent building blocks of sound, I could create a sound system that used almost only English sounds, but sounded very un-English—and as monstery as possible. I called the language Ur-Hag Hesh (literally, Not-Food Language), but I'll just call it Warig here.

We can have a look at some of the important building blocks of syntax in human languages, by looking at some of the properties of my invented language. Though I made it up for monsters, I used my knowledge of the linguistics of natural human languages as a blueprint.

Have a look at a translation of one of the lines from the script for the show, spoken by one of these monstrous creatures, after our hero has desecrated some sacred burial grounds—I think this line ended up on the cutting room floor, but it has all the bits and pieces I need. A Warig priestess says: 'See that the ghosts can rest again'. In Warig, this works out as the following sentence:

Dzhe kishkitaz rul tshakh u-gash, ha u-shak.

There are no consonant sounds made with the lips here. You can say the whole thing with your lips pulled back, baring your teeth—try it! The *dzh* sound is like the 'j' in 'joke', and the *tsh* sound is like the 'ch' in church, and they are formed by combining the basic sounds *d* and *zh* together (and *t* and *sh*). The *kh* is like the 'ch' in the Scottish pronunciation of 'loch'. If you read out this sentence, it sounds very un-English, but it's also quite pronounceable—at least for actors who can do a Scottish accent.

There's also a very particular grammar to Warig, which causes certain sounds to repeat themselves and the recurring patterns of sound and of grammar ring bells in our heads, bells that say: this is a language, it's real, it's not just random sounds. The idea is that, as more and more of the language is heard, that impression gets stronger in the minds of the watchers of the show. Believability and authenticity come from our ability to subconsciously pick up on the language-like patterns we hear: our sense of structure at play. That's the reason that the TV company wanted a made-up language, rather than just coming up with random sounds for their creatures.

I took the basic building blocks of sounds available to human language, and selected a particular subset, an organized and coherent subset, to create the sounds of Warig. These various sounds are not meaningful in themselves. But we can put together these meaningless bits and pieces of language to create meaningful ones.

Think of the word *cat*. The sounds that go into making up this word don't have any particular meaning as part of the word. The *k* sound that begins it is meaningless, and though *at* which ends it does have a meaning in English, it doesn't have that meaning inside this particular word. The word *cat* is a pattern of sounds, and as a whole pattern, it has a meaning, bringing to mind a furry miaowing mammal. But the fragments that build the pattern are themselves meaningless, like the notes of a melody, or a bird's song.

Now think of *cats* (the word, not the felines). This word can be broken up into two smaller patterns of sound: *cat* and *s*. In this word, each of these has its own meaning. *Cat* is what gives you the furry miaowing creature, and *s* tells you that there's more than one of them. So this word is a larger pattern, built up out of smaller meaningful patterns. The larger pattern has its meaning in a way that depends on the meanings of the smaller patterns. If we put that word next to another, building up *Cats miaow*, we get yet more meaning—a whole sentence that communicates something about the world.

If we go back to my Warig example, there are patterns within patterns there. Let me repeat it:

Dzhe kishkitaz rul tshakh u-gash, ha u-shak.

Remember that it means 'See that the ghosts can sleep again'. In Warig, this meaning is expressed a little differently from English. The basic word for conveying the meaning of 'ghost' is *kishkit* and for 'ghosts' is *kishkitaz*. So it seems that the Warig *az* is just like English *s*. But things are actually more interesting than this. Let's look at the Warig for 'See that the ghost can sleep again', where we're just talking about one ghost:

Dze kishkituz rul tshak u-gash, ha u-shak.

Hmm. One ghost is actually *kishkituz*. And something has changed with the word *tshakh*. It's become *tshak*. It gets even less like English when we look at the Warig for 'See that the two ghosts can sleep again'.

Dzhe kishkitukh rul tshaka u-gash, ha u-shak.

More changes. Now the word for ghosts is *kishkitukh*. I designed Warig so that it marks the distinction not just between one thing and more than one, but between one thing, two things, and more than two. There's no special word in the Warig sentence for the concept 'two', unlike in English. The concept is expressed by a particular pattern of sounds in the word for 'ghost'.

We can put this information in a table, like this:

	one	two	more than two
ghost	*kishkituz*	*kishkitukh*	*kishkitaz*

As well as the new word pattern *kishkitukh*, once again, there's a change later in the sentence. We get *tshaka*, rather than *tshakh*. This consistent patterning of sounds, linked across words in sentences in meaningful ways, again contributes to the sense that Warig is an authentic language, and not just a random collection of sounds.

What is this word that variously appears as *tshakh*, *tshak*, and *tshaka*? It's basically like the English word 'can'. It signifies the capacity of the ghosts to sleep, or the possibility that they will. And it changes in a way that is parasitic on the changes in the Warig word for 'ghost'. As we vary the number of ghosts, this word changes its form to track that. We've seen this already in English. The verb agrees with the Subject:

The cat miaows.

The cats miaow.

The word 'miaow' is a verb, and it changes its form to track how many cats are miaowing. The Warig word for 'can' similarly changes form depending on its Subject. Warig also has Agreement. This gives us:

agreeing with: can	one	two	more than two
	tshak	*tshaka*	*tshakh*

The patterning of words we have just seen, distinguishing between one thing, two things, and more than two things, is actually not uncommon in natural human languages. It's a building block that's used in many languages—in fact the ancestor languages of English used to make this distinction. The decision I made to put this kind of system in the language of the Warigs means that it expresses things a bit differently than English. The grammar of Warig is sensitive to a distinction that the grammar of English doesn't care about.

I put this system into Warig really because I was feeling a bit nerdy and wanted to do something different, to make it interesting. But since it's a basic design principle it runs through the whole of Warig. This became rather annoying for the producers. When they asked for the translation of 'See that the ghosts can sleep again', I needed to ask how many people this order was being given to. This is because the distinction between one, two, and more than two works its way into all the nooks and crannies of the grammar of Warig. The producers told me that the order was to just one character in the scene. If it had been two, my translation would have been:

Dzhe kishkitaz rul tshakh u-gash, ksha u-shak.

The change from *ha* to *ksha* in Warig is a pattern that marks whether an order is being given to one, or two people. If it's more than two, we get *za* rather than *ha*.

Dzhe kishkitaz rul tshakh u-gash, za u-shak.

We can have a new table showing this:

number of people ordered:	one	two	more than two
	ha	ksha	za

The English sentence 'See that the ghosts can sleep again' doesn't make these distinctions, and English in general doesn't use the shape of words to distinguish two from more than two things.

Warig does do this and does it in spades. In fact, the language makes many distinctions that English doesn't. In the end, the producers sent me whole scripts rather than just single sentences. This meant that I could see the context for myself. I could work out how a Warig would say the sentence. This is quite different from producing a word for word translation of the English.

I said earlier that the basic word for the concept of 'ghost' is *kishkit*, but we haven't seen a sentence with this basic word in it yet. So do we ever get just *kishkit*? Yes. The translation for 'kill the ghost'—if such a thing is possible—is:

Kishkit ha u-zkat.

We see, again, the word used to make the sentence into an order, *ha*, so we know that just one person is being ordered to kill the ghost. The word for ghost is just *kishkit*, and not *kishkituz*. Why is that? We're still talking about one ghost, after all.

The difference is that the ghost is not the Subject of the sentence anymore: very roughly, it's being acted upon (it's being killed) rather than, at least potentially, doing something (sleeping). It's the grammatical Object, rather than the Subject. Objects, like Subjects, are not really defined by their meaning ('person acted upon'): we saw in a previous chapter that Subject is a purely grammatical concept, defined by the grammar of the language. The same is true for the concept of Object.

Warig changes the forms of its words depending on whether they are Subject or Object. This isn't too strange. English does the same. After all, we say:

She saw the cat.

but:

The cat saw her.

There's a sense in which *she* and *her* here are the same word. They both pick out a female person and just one such person. The only difference between the two is grammatical. We use *she* in English when the word for a single female is a grammatical Subject, and *her* when that word is a grammatical Object. This is a purely grammatical, as opposed to meaning related, difference.

Warig is a bit more oomphy in how it expresses these grammatical distinctions than English: it cross-cuts the Subject/Object distinction with the rich distinctions it makes in terms of the number of ghosts being referred to. This means that the English sentence 'Kill the ghosts' in Warig has six different translations, depending on whether one (*ha*), two (*ksha*), or more than two (*za*) people are being ordered to kill two ghosts (e.g. *Kishkituk ha u-zkat*), or more (e.g. *Kishkitat ha u-zkat*): the words change their form, their sound, depending on what their grammatical role in the sentence is.

Let's go back to the original sentence one last time and see what's going on in the rest of it:

Dzhe kishkitaz rul tshakh u-gash, ha u-shak.

The last few bits of the Warig sentence that we haven't looked at yet are the word *rul*, which means 'again', the word *u-gash*, which means 'sleep', *u-shak*, which means 'make', and the little word *dzhe* which starts off the whole thing, and means something like 'so that'. So literally, the whole sentence is:

So that ghosts again can sleep, make!

We can line up this kind of semi-English with the Warig, so we can see at a glance, which bit of Warig corresponds to the meaning expressed:

Dzhe	*kishkitaz*	*rul*	*tshakh*	*u-gash,*	*ha*	*u-shak*
So-that	ghosts	again	can	sleep,	??	make

We need to put in something to line up with the word *ha* in the Warig sentence to mark the fact that we use *ha* to order one person to do something, not two or more. Unlike, say, *rul*, which has a straightforward counterpart in English 'again', nothing in English corresponds to *ha*. But we need to explain what that word is doing in Warig.

Ha is made up of basic building blocks found in human language in general. It does three different things. First, it expresses a command or order. Second, *ha* also expresses that the command is aimed at just one person. We've come across the idea of grammatical Number before. The two most common grammatical Numbers are Singular and Plural. Just like the notions of Subject and Object, these grammatical concepts don't map on perfectly to meanings. For example, in my bathroom I have a single object. It's flat, and you stand on it, and when you do that it tells you what your weight is. The object is scales. These scales are pretty useful if I want to check whether my cat Lilly has put on weight. Hmm. It's a single object, but I just used a Plural noun (*scales*) with a Plural verb (*are*) in that last sentence to talk about that single object. This means that the word is grammatically Plural, even though it's used to talk about a single thing. There are good historical reasons for this—not so long ago, scales used to have two clear parts to them, like trousers and scissors still do—and those historical reasons are why English treats that word as Plural. But nowadays, the word isn't usually used to refer to something that is Plural—unless you have a gigantic set of scales and weights in your bathroom, and you balance on these each morning. Grammatical number has come apart from the meaning of plurality. Grammar is linked to meaning, but it isn't perfectly aligned with it.

The third and final notion that *ha* expresses is that the command is aimed at whoever the utterance is being addressed to. This might seem like common sense—if you are commanding someone to do something, it's usually the person you are talking to—but actually languages sometimes have special forms to command someone else than whoever is being addressed. For example, in Kumyk, a Turkic language spoken in Dagestan, North Ossetia, and Chechnya, to tell someone to go, you say *Bar!*, but to express that another person should go, you use the form *Barsin!*, 'Let him/her go!'. The basic building block of language that signifies that a statement is being addressed to someone is known as the second Person (where the first Person is who is speaking, and the third Person is someone else). Linguists usually abbreviate first, second, and third Person as just 1, 2, and 3.

So with these general building blocks of language in mind, we can line up everything like this:

Dzhe	kishkitaz	rul	tshakh	u-gash,	ha	u-shak
So-that	ghost.PL	again	can.PL	sleep,	IMP.2.SG	make

When I sent my translations to the producers, I annotated them as you see here. I immediately got a question back about what the strange numbers were, and had to explain they were just for my own use, so I could work out the right patterns to keep everything consistent. But it does look like you're doing some weird maths.

The little abbreviations sprinkled over the semi-English version mark aspects of the grammar of Warig using some building blocks of language that linguists have discovered. You can guess that IMP means imperative (an order), that SG means Singular, and that 2 means second Person. And the PL that appears after 'ghost' and 'can' marks that they express Plural (more than two, in Warig).

You can see how helpful this system is. We've used a simplified version of it in previous chapters, but now you get to see it in its full glory. It's called glossing: we use a rough English translation, line it up with the language we are translating, and add in elements from the basic building blocks of human language that show how the grammar of the particular language we're interested in works. There are special glossing rules that linguists use so they can be consistent with each other. Do an internet search for Leipzig Glossing Rules if you're interested. The fact that there are less than one hundred basic building blocks in these glossing rules begins to hint at just how restricted the categories of human language are. Even if we multiply that number by ten, there will still be less than a thousand basic grammatical categories of human language—in my own view there is actually a much much smaller number. From this small number of basic grammatical building blocks, all the words in all the languages of the world are created.

The way I designed Warig is actually nowhere near as weird as human languages can be. Some languages also have special forms to express that there are three things being talked about, and some even go further than this and make differences between one thing, two things, three things, a small number that's bigger than three, and a larger number bigger than that. Languages like Yimas, and Fijian are examples of languages that are very rich in how they express numbers of things. So Warig is pretty conservative actually. The building blocks I used for making it up weren't too wild in the end.

Imagine, though, I'd decided that Warig would have special forms for the following: one form for when we are talking about one to four or six to nine things, a different form for exactly five or ten things, and a final form for more than ten things. Let's call this the hand-number system. Our new version of Warig would have the following table:

	5 or 10	1–4 or 6–9	more than 10
ghost	kishkituz	kishkitukh	kishkitaz

No human language (we know of) has a grammatical way of marking number that works like this. This is even though we have five fingers on one hand and ten fingers altogether on both of our hands. So it might be a system that could be useful to us, or the Warigs, who as far as I know, have the same number of fingers as humans. We'd have one form of the noun for when you use a finger to count, another for when you use a full hand (or maybe fist), and a final form for things that can't be counted on our hands. These different categories of grammatical number would be active in the grammar, so that the noun would change form like it does for one, two, and more than two in Warig, and maybe the verb would agree. But if we look at the human languages we know about, there's a gap. We're just missing languages that look like that.

Here's another kind of system that's missing, even though all the relevant distinctions are made by languages we do know about. Imagine a language where there are three patterns for expressing how many things you are talking about, just like in Warig. But where Warig, and many human languages, make a distinction between one, two, and many, this imaginary language would have a form of words for one thing, a different form of words for two or more than three things, and a special form for just exactly three things. There are still three forms for the words; the forms just pick out different numbers of things: a special form for one thing, a special form for three things, and another form for either two or more than three things. This pattern is also absent from the world's languages, even though languages like Yimas and Marshallese can happily express 1, 2, 3, a few, and more than a few in their grammar, with the verb tracking the grammatical number of the nouns in the sentence.[2]

In fact, we're missing a lot of very plausible languages. All 7,000 human languages we know of either don't use number at all, or they have systems that are based on concepts of one, two, three, a few, and many.

Why are all the other patterns missing? There are some different possibilities. It could be an accident, pure historical chance. Perhaps there was a language spoken in the distant depths of history by the first ever speakers, and they just happened to use that system of grammatical number, which has then been passed down through the generations? It could also be because as languages change, they only keep around those patterns that are useful or perhaps common. We'd need to work out exactly what we mean by useful, but perhaps it could be to do with what is culturally useful. Another possibility is that when children (or little Warigs) learn languages, the way their learning works biases them against certain patterns, or even makes certain patterns impossible to learn. Their minds are limited in certain ways so that some logically possible languages are psychologically impossible. And it could be a mixture of all of these things.

When I designed Warig, I didn't go too exotic. I kept to basic linguistic building blocks and systems for combining them that are used in many natural human languages. But I could have gone crazy, and broken the rules dramatically. Let me do this now and invent a new language, or a very small bit of one. I'll then use this to explore the various ideas I just mentioned.

In Warig, the Object of a verb comes before it. We saw this earlier, with the translation of 'Kill the ghost'. In the Warig sentence, the word for ghost *kishkit*, comes before the verb for 'kill' with its marker of command:

Kishkit   ha   u-zkat!
ghost   IMP.2.SG   kill

This is the opposite order from English, where the Object ('ghost') comes after the verb ('kill'):

Kill the ghost!

Now, we've seen that the sounds of Warig split up into certain classes, ones made with the glottis, ones made with the back of the mouth, and ones made with the tongue and the teeth. So let's make up a different language, which works like Warig when the Object begins with a back of the mouth sound, and like English when it begins with a toothy sound, like the word for 'priest', which is *tshul*. This weird Warig language would then translate 'Kill the ghost!' just as we've already seen, but 'Kill the priest!' would be *Ha u-zkat tshul!*:

ha	*u-zkat*	*tshul*
IMP.2.SG	kill	priest

This rule makes the order of words in this new version of Warig depend on what sound those words start with.

This is an easy rule to apply. Quite a bit easier than the complex system for number, but there is not a single natural human language ever discovered that works this way for ordering words. The order of the verb and the Object in human languages never depends on the sound that the Object starts with. In fact, the same is true for Subjects and verbs, or for the relationship between a Subject and an Object. The sounds that words start with, or end with, never play a role in the basic rules of word order in a language. Could this just be a historical accident that happens to have affected every language ever looked at? It seems unlikely.

Or here's another example. We can make vowels longer or shorter in English, and we do this in some words to signify emphasis, or sarcasm. Imagine someone invites you to a talk on theoretical syntax, and (curiously) you're not that interested in listening to it. You might respond:

Yeah, cos I'm soooooo fascinated by syntax.

We use this variation of vowel length in English to express attitudes in various ways. Now English also expresses Tense on verbs. The difference between 'I love syntax' and 'I loved syntax' is that in the second sentence, the loving took place in the past, that is, sometime before the sentence is said. How long ago something is in the past is continuous: earlier today, yesterday evening, yesterday afternoon, yesterday morning, and so on.

In fact there are languages where different degrees of something happening in the past are grammatically marked. Yimas, the New Guinea language we already met, distinguishes between events that happened yesterday versus those that happened sometime between the remote mythological past, and about six days ago. So Yimas has a recent past, a far past, and a remote past Tense, all marked on its verb.

So now imagine a new language, where we vary the length of the vowel like we do in English to express sarcasm, but we connect that to how long ago the event we're talking about took place. This language would be like English, but we'd have translations like this:

I love syntax.	↔	*I love syntax.*
I loove syntax.	↔	*I loved syntax earlier today.*
I looove syntax.	↔	*I loved syntax last night.*
I loooove syntax.	↔	*I loved syntax yesterday evening.*
...	↔	...

This seems like quite an easy rule to use, or learn, but we have never discovered a human language that does this. Much more generally, we don't know of languages that link continuous variation in sounds with continuous variation in the meaning of the grammatical expression. We don't find languages that mark how much agency someone has in carrying out an action (something which can vary continuously) by increasing or decreasing the pitch of the vowel of the verb. We'll see, in Chapter 6, that the syntax of human languages works only on discrete units of language. Continuous aspects, even though we humans can perceive them, are invisible to syntax and the way that syntax builds up meaning.

This is interestingly different to what happens with certain concepts connected to attitude, identity, and other social factors to do with how a speaker presents themselves while speaking, as in the sarcasm example. These can be linked to continuous variation in how sounds are produced, or to the frequency with which certain grammatical markers are used, as we'll see in Chapter 10. But as far as grammatical concepts are concerned, language treats these as discrete, not continuous.

Here's one final example. In English (and in Warig), as we've seen, there is grammatical Agreement. The verb changes in concert with the Subject. So when we have just one cat, in the following sentences, we get 'miaow', but more than one cat gives us 'miaows':

The **cat** miaows.

The **cats** miaow.

The rule for when you get this change in the verb is not as simple as you might think. The easiest rule would be to say that the form of the verb depends on the word that comes immediately before it. But that simple rule can't be right, since you can put things between *cat* and the verb. You can say *The cat always miaows*, so the rule can't literally care about the immediately preceding word. Maybe it just cares about some preceding word that expresses a number of things—in English, whether there's one or more than one. But actually, that won't work either, because of sentences like this:

The **cat** under the trees miaows.

In this sentence, the word *trees* is Plural—there's more than one tree. But the verb is still working in concert with *cat*, and doesn't seem to care about the word *trees*.

We could easily design a language, however, that did work in this way. Such a language would have the following kind of rule:

Make sure that the verb changes in form in a way that is influenced by the closest word expressing Number before it.

In a language like that, we'd say the following:

The **cat** miaows.

The **cats** miaow.

The cat under the **trees** miaow.

The cats under the **tree** miaows.

In these examples, the verb shows Agreement, not with *cat(s)*, but with whatever the immediately preceding noun is.

This is a very easy rule to write, or to follow, and I could have designed Warig to work like this with no problems. In fact, if I had, it would probably have still felt and sounded language-like to people hearing it. But it's not a rule that English, or any other language we know of, follows. This is very strange, because it seems the most obvious kind of rule. You can imagine children making this kind of mistake and keeping it going as they grow up. In actual practice, there wouldn't be huge numbers of examples where things would go wrong, as, pretty commonly, the relevant noun and its verb are, in fact, right next to each other. If that were to happen, we would see the language changing slowly until it eventually adopts a new rule, one that depends on the closest noun to the verb, not on what actual Subject of the verb is. But, again as far as we know, that has never happened. There's a gap.[3]

When we look at languages, we see the same concepts being co-opted into grammar over and over again. These are the basic building blocks of language. Languages can look very different on the surface—after all, they sound pretty different—but deep down they work via very similar principles, the Laws of Language applying to these basic building blocks.

One of the most important Laws of human language is that grammatical rules refer to structures, like the ones in the pot-dealer story that we used to distinguish the meanings of *public park or playground,* or like the structures we saw emerging in David's Homesign in the previous chapter. We've just seen that no language that has been studied so far has Agreement between a verb and the previous word. What languages care about instead is the abstract structure—the grammatical Number of the Subject of the sentence, in this kind of example.

There's also interesting neurological evidence that backs up this general point about the importance of structure. A team of neuroscientists in Germany and Italy led by Maria-Cristina Musso and Andrea Moro carried out a study which was published in the scientific journal *Nature* in 2003.[4] The team took speakers of German and wired up their brains to MRI machines. The scientists then taught the German speakers some parts of languages that they had no knowledge of at all (Italian and Japanese). The twist was that some of the grammatical rules they taught the speakers were real rules of the new language and others were made-up 'impossible' rules.

An example of a real rule is the rule to change so called 'active' sentences into 'passive' ones. An active sentence is just the kind of structure we've been talking about a lot so far: In a sentence like *The poet wrote the poem,* you have a Subject *the poet,* a verb *wrote,* and an Object *the poem.* In the passive version, you make the Object into the Subject, and what was the Subject ends up being accompanied by the little word *by* (in English). So we get *The poem was written by the poet.* The process is quite similar in German, but there you use a special verb which roughly means 'become', and the little word analogous to *by* in English is one that means 'from'. So, lining up the German with a rough translation in English, we get:

Das	*Gedicht*	*wird*	*vom*	*Dichter*	*geschrieben*
the	poem	became	from-the	Poet	written

The order of the words is a bit different in German (the verb comes at the end, for example) but the general process is pretty similar.

Now, Japanese does something completely different from this. In Japanese to get the passive, you change the ending of the verb, adding on the affix *reru*. In Japanese, the sentence is:

Shi	*wa*	*sakka ni*	*kaka-reru*
poem	SUBJ	poet by	written-PASSIVE

In this sentence, the little word *wa* appears immediately after the Subject of the sentence. Both languages make the Object into the Subject in a passive. In Japanese, the difference between an active and passive boils down to which noun is followed by the little word *wa* and to what the ending of the verb is. In German and English, it's about the order of the nouns (the Object appears where the Subject usually would be) and there's a separate verb that signals that the sentence is passive.

The researchers taught the German speakers the Japanese Passive rule, but they also taught them another kind of rule, one that pays no attention to the structure of the sentence. The researchers made up a rule for pretend-Japanese where to make a sentence negative, you put the negative word *nai* after the third word of the sentence. This rule doesn't care about structure. It just cares about the order of words in a sentence, as though the words in the sentence are laid out like a set of beads on a string. For example, using this 'impossible' rule, the negative of the sentence above would be:

Shi	*wa*	*sakka*	*nai*	*ni*	*kaka-reru*
poem	SUBJ	poet	not	by	written-was

Here, we just count the number of words from the left of the sentence and put the negative word *nai* in after the third word. This isn't a rule of Japanese, or of any known language. If rules of human languages care about abstract structure, as most linguists think, then this would be an impossible rule for language. The researchers tested out the passive rule for Japanese and the impossible negative rule across hundreds of sentences with their German speakers and analysed the results of the brain scans.

They found a striking difference. The German speakers were able to learn both the real and impossible rules with some accuracy. They were a bit better with the real rules. But different parts of their brains began to get involved as they gained more and more practice using the rules. It was as though the speakers' brains treated the two different rules as quite different kinds of thing.

There's a special area of the brain known as Broca's area—if you put your finger just above and to the front of your left ear, it's round about there. For over a century, Broca's area has been thought to be involved with language, mainly based on studies of people with brain damage and language problems. It was a subpart of Broca's area that became more and more active in the brain scans of the German speakers as they used the real rules, but this didn't happen when they used the impossible rules. The brains of the German speakers reacted differently when they were learning a possible—indeed actual—rule of a human language previously unknown to them, versus a rule that ignores linguistic structure and pays attention to just the pure order of words.

It's important to be careful when drawing conclusions about experiments like this, especially since the participants were adults who already have a language, and they were given explicit instructions, unlike children who come to their first language cold. At a minimum, though, the experiment shows that our brains are attuned differently to the kinds of linguistic structure linguists find in languages from those that we don't find. Linguistic research tells us that there are building blocks of language, and these are put together in particular ways, crucially involving syntactic structure. The neuroscientists' experiments tell us that human brains are sensitive to that structure.

So far in this chapter, I used my experience in designing the language of the Warigs as a way to show how human languages can be put together, and how they can't. I gave a range of examples of made-up languages that based word order on the sounds of the words, that based continuous changes in grammatical meaning on continuous changes in sounds, and that based grammatical rules on the closeness of words to each other in a sentence rather than on their structure. These seem like reasonable rules, but we never find them. More than that, it appears that our brains are not well set up to use certain rules that we don't find in actual languages.

In designing the language of the Warigs I also elected to use grammatical concepts that are well established in languages of the world. We saw a few of these. Words in Warig change their form depending on the number of things they refer to: one, two, or more than two. They also change depending on their grammatical role in a sentence—whether they are Subjects, or Objects. As we've seen from the Warig examples, there is also a special grammatical concept for giving orders (the imperative) and for marking who is speaking, who is being spoken to, and who is being spoken about.

A big question in linguistics is: What grammatical concepts exist in the world's languages? I said that we don't know of any language where grammatical Number works on the hand-number system, even though such a system doesn't seem implausible. There is a vast number of possible concepts that logically could appear in the grammar of a language. Some of these are very implausible. For example, you could easily make up a language where the verb takes on a special form depending on whether it's currently raining versus sunny, versus snowing; or where a noun has an affix that signals whether the person being referred to is allergic to nuts. In all of the centuries of work on languages, no one has ever found languages that do either of these things. Languages can easily use sentences to say whether it's raining or snowing, or whether someone is allergic to nuts, but they don't change the form of verbs to do this. These are not concepts that human grammar expresses.

On the other hand, languages do quite typically mark verbs for whether or not the event or situation the verb describes takes place before the person speaking utters the sentence. If I say 'I love coffee' the loving is happening while I'm speaking; if I say 'I loved coffee', whenever that loving happened, it's no longer going on while I'm speaking. The verb changes its Tense depending on whether the situation being talked about happened before the situation where the sentence is said.

There's a contrast then. Although there are many languages that mark a verb for Tense (like 'kick' versus 'kicked'), there aren't any languages where I can say:

I kickuz the door.

meaning, 'I'm kicking the door and it's snowing'. To do that, I have to say 'I'm kicking the door and it's snowing'.

There are also more plausible cases of things that might be grammatical concepts, that still don't appear to occur. No language we have discovered marks verbs for whether the speaker is pleased about the situation, or thinks they have been in that same situation before, or that it might be dangerous, or advantageous to mating, or a host of other possibilities. Scientists have shown that our visual system is acutely sensitive to whether what we are looking at is symmetrical around a central axis or not. But no language we have ever discovered marks nouns grammatically for whether they are used to refer to symmetrical objects (like human faces, the sun, or certain flowers) or asymmetrical ones (like coastlines, star constellations, or other flowers).

That's not to say that all languages express just the same concepts that are expressed in English. We already saw that some languages distinguish not just between one and more than one, but between one, two, and more than two. There are also languages that mark a verb for whether the speaker, for example, got their information about the situation second-hand. For example, in the Native American language Cherokee, which is spoken in North Carolina and Oklahoma, and is part of the Iroquoian language family, you simply have to express, on the verb, whether you got the information about the situation you are describing first-hand or whether you heard it from someone else, or are inferring it in some other way.[5] You can't just say 'It rained' in Cherokee, you have to say either:

U-gahnan-d'i.

or:

U-gahnan-e'i.

The first of these means that it rained and you know that because you saw or heard the rain. You'd use the second if, for example, you woke up and saw everything outside was wet. Linguists call this Evidentiality, and though English doesn't express this concept obligatorily in a sentence, many languages do.

Korean, for example, works very like Cherokee, although the two languages are not remotely related to each other. If a Korean speaker wants to say that their friend Toli ate some dumplings, they have to say either:

Tolika mantwulul mekess-e.

or:

Tolika mantwulul mekess-tay.

Notice that little change at the end of the sentence: either -*e* or -*tay*.[6] The first one signals that the speaker has actually seen Toli eat (*mekess*) the dumplings (*mantwulul*). The second sentence indicates that they have just inferred that Toli ate the dumplings. They didn't see it happen, but perhaps Toli just looks very satisfied.

Similarly, some languages have a special marking on the verb that signifies whether the speaker is surprised by the event that they're describing. This is called a Mirative, from the Latin word *mirare*, which means to marvel or be astonished—we see this same root in the English word *admire*. When languages use Miratives, they use them to signify a sudden change in the speaker's knowledge, from not knowing something, to knowing it. This often leads to a sense that the speaker has just realized something, or is surprised by it.

Chechen, a language of the Northeast Caucasus, has a special marking on its verbs that signifies that the speaker is reporting an unexpected event. For example, imagine that you and I are Chechen speakers and we watch some children playing with a chicken. The children try to put the chicken into a cage and it goes in, but suddenly it gets back out again. I could turn to you and say:

hwazhahw, jiediq isa

The whole thing means 'Look! It's escaped!'. The first word is a polite way, in Chechen, of saying 'Look!', while the word *isa* just means 'it'. We're interested most in the verb, *jiediq*, which means 'has run' but crucially also marks, via the *iq* at the end, that I wasn't quite expecting the escape to happen. I thought the chicken was safely in the cage, and I've suddenly realized that it's not.

Lots of languages encode Mirativity in their grammar. Chechen is just one case. We also find Mirativity in Tibetan, in Quechua (a language spoken in South America), in Native North American Languages, in the !Xuun languages spoken in Namibia, Botswana, and Angola, and elsewhere. In fact, there's some grammatical evidence for Miratives in English too.[7]

In English we can tell someone about the location of something by saying:

It's here.

Now this sentence can convey a sense of surprise, or discovery— perhaps you just found whatever it was you were talking about— but it doesn't have to.

But in English we can also say:

Here it is!

This exclamation, in contrast must signify that you've suddenly discovered the location of something. For example, imagine I have a party, and everyone puts their coats in the upstairs bedroom. You're tired, and want to leave, so we both go to that room to get your coat. Without finding it, I can still say:

It's here. It's somewhere in this room.

But it's really weird to say:

Here it is. It's somewhere in this room.

To say *Here it is* you need to have just discovered it. You mark, using this special word order, that you've gone from not knowing where the coat is, to knowing where it is. This is the hallmark of Mirativity as we see it elsewhere. The speaker goes from not knowing what is being talked about, to suddenly knowing it.

Mirative grammatical marking, then, appears widely in unrelated languages. It even crops up in those languages which, like English, don't make widespread use of it. The same is true for Evidential marking. These are both common building blocks of language.

Contrast this with what we might call, if it existed, sollicitative marking, from the Latin *sollicitare*, meaning 'to worry or be anxious about something'. No language we've ever discovered has a grammatical marking for this, though worry or anxiety would seem to be as pervasive aspects of human experiences as surprise.

Surprisingly, it appears that there are a limited number of truly grammatical concepts, that appear over and over again in the world's languages. The flip-side of this is that there is a huge number of perfectly fine concepts which are never co-opted into the grammar of human languages. Grammar has its limits.

A human language is not just a random collection of sounds. Meaningless sounds are organized into meaningful patterns—roughly the words of the language. One important aspect of many words is that they are symbols, directly linking an idea or concept with a sound. But words are much more than this. Words, when they are used as part of a language, are connected to the basic building blocks of human language itself. Grammatical concepts like Number and Tense, Subject and Object, noun and verb, are used to categorize and alter words so that they can be combined into sentences. It's this sentence building capacity that allows human language to be so limitless in scope. At the same time, the restricted number of grammatical concepts curbs the possibilities for human language. The number of grammatically realized concepts is tiny compared to the huge range of psychologically possible concepts.

As well as the finite number of grammatical concepts, the grammatical relationships between words in human languages are also extremely restricted. One might imagine that the most obvious relationship between words—whether one word is said right next to another—would be the one that is most widespread in human languages. But this is *never* found as a fundamental organizing principle in a language. Human languages instead use an invisible structure to connect words that might be spoken quite far apart. Language is highly constrained, both in what grammatical concepts it uses, and the ways these can be connected to each other.

Languages do use their grammar to mark certain concepts that seem exotic to a speaker of English. However, what we find, as we look across languages, is that the same concepts crop up again and again encoded into grammar—it's rare that linguists find a new grammatical category that we're unfamiliar with. We can only investigate the languages we have, but it appears that they converge on a limited list of concepts that language, as a whole, cares about enough to encode in grammar. Not every possible concept is encoded, and those that are encoded appear over and over in very unrelated languages.

Not every category of thought is a possible category of language and not all possible systems of symbols are possible human languages. There are limits, imposed by the human mind, on linguistic experience.

6

ALL IN THE MIND

U sually we humans are very proud of our ability to perceive the world around us, to know what is real, and to act upon it. These perceptions, together with our intelligence and our ability to solve problems and understand the world around us in creative and flexible ways, are what make us so successful as a species. But it turns out that certain aspects of the way we structure our experiences are actually forced upon us by our limitations. Sometimes these limitations cause us to perceive things that just aren't there, like the coiling arms of a spiral in Figure 4.

This illusion is called the Fraser Spiral. No matter how hard you look, the lines here seem to trace a spiral. But they don't. If you put your finger on one of the arms of the spiral and you trace it round, you'll see it's actually a circle. The reason we interpret this how we do is because of the way we are constituted as human beings. This constitution is just as much part of who we are, of human nature, as our intelligence and creativity.

The evidence of our senses is processed, organized, analysed before it gets anywhere near what we think of as perception. One of the most important things that our brains do is chop our experiences into discrete chunks. These chunks aren't there in

Figure 4 The Fraser Spiral

the noise that surrounds us: we create them. Our experience of language is a continuous stream of sounds or body movements, but we analyse it as a structure made out of of discrete linguistic atoms, combining in various ways. We do this with the basic units of sound. We turn combinations of these into words with particular meanings. How do we get from the complexity of experience to these building blocks of sound and of sentences? And are we the only species that can do this?

Say the following sentence out loud:

Lilly scratched Anita.

The sounds of the sentence you just said, if you measure them using acoustic equipment, are continuous waves of vibrations in the air. We can display this visually, using a computer programme to produce what linguists call a spectrogram, as you can see in Figure 5.[1]

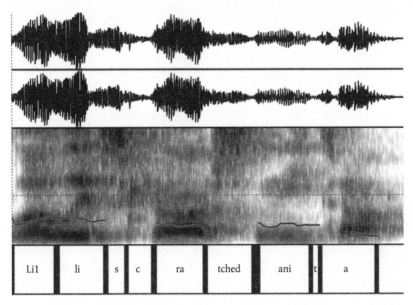

Figure 5 A spectrogram for *Lilly scratched Anita*

This spectrogram displays various aspects of the noises made when my nephew Sam said *Lilly scratched Anita*. The top part of the diagram represents, very roughly, how loud the noises are (how much pressure the sound has). The bottom gives us information about the frequencies of the noises, their pitch. As you can see, the diagram is continuous. Where the words start and end is not where the various sound pulses start and end. This visual representation tells us that sound waves aren't discrete.

That, however, is not how we perceive the sentence. What we actually perceive is abrupt changes in the sounds. We have a sense that there are two distinct 'l' sounds when you say the word *Lilly* (and of course there are three letters in this spelling of the word), one at the start of each syllable. But in the spectrogram, all we see are continuous changes of frequencies and noise level.

There's a disconnect between what is out there in the world when we use scientific equipment to measure it, and how we mentally interpret the signals that our senses pick up.

This fact about speech has been known for over half a century. Experiments, beginning in the late 1950s, then developed over the following decades, showed that, if a syllable like *ba* is played to a speaker of English, and is manipulated by computer so that the properties of the sound are gradually and continuously changed to *da*, we do not perceive a gradual change.[2] Rather what we hear is a sharp change from a *b* sound to a *d* sound. Even though the actual physical noise that is played involves gradual change, humans don't perceive that gradualness. Instead, we impose a sharp boundary between the two classes of sound. Our language works with discrete categories, rather than with continuous ones. This is true whether the people doing the experiment have a literate culture or not, so it's not connected to exposure to a spelling system that involves an alphabet. The experiment also works for speakers of languages whose writing system doesn't use discrete units of sound as a basis, for example people who speak and write Mandarin Chinese. It also works for babies, who haven't been exposed to any writing system at all. Most amazingly, it seems to work in the same way for babies across different languages, even if those languages don't make exactly the same distinctions as English.[3] This phenomenon is called categorical perception.

Also in the late 1950s, Noam Chomsky had begun to make arguments that human linguistic abilities are not part of our general learning skills but rather are a result of specialized innate capacity, a hugely controversial idea at the time. The ability of even very young babies to process continuous sounds as discrete units seemed to be evidence that Chomsky's idea was true not just of syntax, but also of something as basic as perceiving sounds.

This idea took hold in the field of speech perception in the early 1970s. More and more experiments showed that babies imposed the same kinds of distinctions on the speech sounds they were exposed to, irrespective of the language of their community. Since speech is such a human phenomenon, scientists initially assumed that these experiments on babies' categorical perception showed a specifically human capacity.

That thought was quickly punctured by a startling study by Patricia Kuhl and James Miller. Kuhl and Miller showed that the ability to cut continuous streams of sound into discrete units was something that chinchillas were also pretty good at.

While the difference between *ba* and *da* is due to where in the mouth the tongue and lips are positioned, sounds which are made in the mouth in exactly the same way can also be distinguished by voicing, as we saw in the last chapter. Take, for instance, the syllables *ton* and *done*. These are extremely similar, with the tongue in the same position in the mouth for both sounds. The difference is in voicing. The vocal cords vibrate during the *d* sound, but don't during the *t* sound.

In fact, this all really comes down to a matter of timing: the vocal cords switch on a little bit earlier as they transit from the consonant to the vowel when you say *done*, than when you say *ton*. The onset of voicing takes longer to happen with the *t*-sound. This is all that really distinguishes *d* and *t*: the timing of when the vocal cords begin to vibrate.

Now it's possible to increment just exactly when the onset of voicing happens in little steps of milliseconds. This can be done using a speech synthesizer. The synthesizer keeps all other aspects of the sound identical, but alters, in a very fine-grained way, exactly when the voicing starts. When scientists play recordings controlled like this, speakers of English, and other languages, report a very abrupt switch from one sound to the other. They impose a sharp boundary, even though the physical acoustic signal that they hear has no such boundary in it. This same technique was used to show that human babies also perceive sharp boundaries between sounds, even when the acoustic signal is altered in a continuous way. Humans perceive speech sounds categorically, even though there are no categories in the actual sounds being listened to.

Kuhl and Miller published a study in the magazine *Science* in 1975 that showed, contrary to what had been assumed, that this wasn't a uniquely human experience.[4] Miller, in some earlier work, had demonstrated that chinchillas and humans have very similar hearing abilities: chinchillas hear in frequencies that match those that humans can hear in. This made chinchillas a great test case for whether animals other than humans had categorical perception. Kuhl and Miller trained four chinchillas to respond differently to speakers saying *t* sounds versus *d* sounds. Then, using a speech synthesizer, they tested whether chinchillas perceived the same sharp, categorial, switch between the two sounds that humans do, a distinction that isn't there in the acoustics, but is imposed by human brains.

Kuhl and Miller did this by exposing the chinchillas to some rather unpleasant experiences. First they held back drinking water from the chinchillas so they were thirsty. They trained them to run across a surface to get to a tube of drinking water. They played sounds to the chinchillas as they ran across the surface and they got the chinchillas to associate pleasant or horrible experiences with the two classes of sound, voiced and unvoiced. For example, if they were training a particular chinchilla to dislike voiced sounds, and like unvoiced ones, they'd set the chinchilla going, play a *da*, and when the chinchilla was running towards the water, they'd electrocute its feet! They'd also play *ta* and reward the chinchilla by allowing it to quench its thirst unmolested. As you might imagine, such a chinchilla would quickly decide that it hated voiced sounds, and loved unvoiced ones. A chinchilla who had been electrocuted every time it heard a very clear *d* learned not to run across the floor to get water if they heard that sound. Kuhl and Miller could then use how the animals behaved to test their perception of the sounds.

Once the chinchillas had learned to love or hate certain consonants, Kuhl and Miller tested to see whether they were sensitive to tiny changes in when the onset of voicing happened. They used synthesized speech to move the timing of voicing forward or backwards in tiny increments. Their question was whether the chinchillas would behave as though there was a sharp shift between the two categories, or whether their behaviour would be more continuous.

The chinchillas behaved almost exactly like the speakers of English—whose feet, it must be said, were not subject to electric shocks: they split the continuous differences into two sharp categories corresponding to the voiced versus unvoiced sounds.

This early work has been replicated again and again with other species. The current consensus is that many species share with humans a capacity to attend to quite subtle properties of the sounds around them. When continuous streams of sounds are heard, many animals, including us, can turn them into discrete categories. We humans have an inbuilt bias to turn what is apparently continuous in the world around us into discrete categories in our minds, but we're not alone in having this ability. In Chapter 3, I gave a quote from the ancient Greek philosopher, Epikharmos of Kos: 'only the mind sees and hears, all else is blind and deaf.' A chinchilla version of Epikharmos would have said the same as the human one: only the mind hears, and it hears in discrete categories.

Is this something we learn to do, or are our brains set up, even before birth, to perceive the discrete categories that are crucial to language?

Mahdi Mahmoudzadeh, Ghislaine Dehaene-Lambertz, and colleagues in France have been studying this and closely related questions over a number of years.[5] They've shown that, even in prematurely born babies, the same special brain areas that adults use to process language are sensitive to distinct categories of sounds. These babies show a specialized and very strong brain response, on the left-hand side of their brains, to the difference between the syllables *ba* and *ga*. This is a pretty subtle distinction in terms of pure acoustics. Because these babies don't even babble yet, this means that our brains are sensitive to distinct categories of speech sounds even when we can't articulate them.

The team also tested whether the babies were sensitive to the difference between male and female voices. This difference is very noticeable when you examine the spectrograms of the physical signal, but it's not part of the sound system of a language.

The babies' brains in this case lit up in a completely different place and in a different way. The language-related areas of the brain barely showed any response. Even though the categories of gender are hugely important in the social structures of human societies, differently gendered voices aren't, it seems, something that the infant language-learning brain systems care about. Rather the brain networks involved in listening to language are pre-tuned to ignore the big acoustic differences between men and women speaking, and to pay close attention to the very subtle signals that distinguish *b* and *g*.

These babies that the team tested were premature. This very strongly suggests that human brains are set up, even before the normal nine months gestation is over, to perceive just the categories that are central to language.

Humans and other animals share categorical perception, the ability to impose discrete boundaries on continuous sound waves. Human brains seem to be set up to not only impose discrete boundaries on sounds, but also to only pay attention to certain properties of sounds. This is similar to what we saw in the last chapter. There, I argued that human languages are restricted in which concepts can enter into grammar. I showed you that continuous aspects of the world never make it into the grammar of languages. Nor do certain concepts. As we saw in Chapter 5, while languages have Miratives, grammatically expressing something like surprise, they don't have Sollicitatives, expressing concern or worry. We have limits, and those limits provide an understanding of why our languages look like they do.

This bias towards using discrete categories holds not just of sounds, but also for other aspects of linguistic structure. Adult speakers of languages easily recognize distinct words, carving them out of the continuous stream of speech. Because of this, if you take a sentence and mix up the words with each other, you can still distinguish them. This is pretty obvious when the words are written down, because in English the tradition is to write spaces between words. But not all languages do this. Putting spaces between words is a fairly late invention in writing systems that use the Latin alphabet; early examples of alphabetic writing divided words using various dots (called interpuncts), or justranallthewordstogetherjustlikethis. As you can see, it's quite possible to separate out the words in a continuous sequence of letters like this, though it's a bit of an effort when you are not used to it.

If you mix up bits of words with each other, however, it becomes much more difficult to work out what is going on.

Compare the following three cases, where the first is a sentence of English; an unusual sentence, but a sentence nonetheless. I've removed all the spaces, but it's fairly easy to distinguish the words:

Tinyelephantssnortchampagnewiththeirtrunks

In the second example, we have the same words, though I've rearranged them so they no longer form a sentence. Again, it's not too hard to pick out the different words:

Tinysnorttrunkswithelephantschampagnetheir

But in the next, and last example, I've rearranged bits of words, rather than whole words. Suddenly it reads as a complete mess:

Tiantselephpagnesnonkschamnywirtthethtruir

Although I've presented these examples in written form, you can easily imagine that you'd find the same effects if you'd heard them rather than read them. The reason that the third example is so hard is because words, as discrete units, are crucial to how we store information or knowledge about the languages we speak.

How do we end up with knowledge of these basic units? When children are born, they don't know the words of their language. However, as they develop, children pick up tens of thousands of words. Researchers who have studied this process report that adults in literate Western societies usually end up with about 60,000 words. If you do the maths, this suggests that children are learning about ten words a day. Part of that learning has to be about pulling out the discrete words from the continuous streams of speech surrounding the child, but there's a lot more than just that. The child has to understand that different pronunciations of the same word by the same or different people are really, somehow, the *same* word. They also have to end up with some sense of the meaning of the word, what grammatical patterns it fits into, and all sorts of other things.

Let's first focus on the most basic part of this: how do children figure out, from the continuous stream of sounds they hear, which parts of that stream correspond to discrete words. The ability of young children to do this is quite remarkable. You might imagine that children learn most words by hearing them one at a time, perhaps carefully taught to them by their parents. But it turns out that less than 10% of words appear on their own. Children learn the vast majority of the words they know by somehow selecting them from continuous streams of speech.[6] The question is how they segment out the words from the continuous stream of speech?

In a now famous work published in 1996, the psychologists Jenny Saffran, Richard Aslin, and Elissa Newport, with various collaborators, have shown what impressive abilities tiny children have.[7] This research team worked with eight-month-old babies. They played the babies streams of nonsense syllables that looked a bit like this:

bidakupadotigolabubidakugolabu...

Though these syllables don't correspond to any particular English words, there are regularities in which syllables come before which others. You can see that the sequence *bidaku* appears at both the start and near the end of the stream of sounds. These streams of sounds were designed so as to have word-like units, where after a particular syllable you'd be more or less likely to get another particular syllable. For example, after *bi*, you'd always get a *da*, but after a *ku*, you'd more rarely get a *pa*.

The researchers played these nonsense syllable streams to the babies using a computer that produces synthesized speech, so they could make sure that there were no hidden cues for the children (extra long pauses, or differences in intonation). What they were interested in was not so much what children actually did when they learned a normal human language, but more in what children *can* do.

It turns out that these babies are pretty good at detecting these variations in how likely one syllable is to follow another in this nonsense language. They are 'intuitive statisticians' and can detect how likely one syllable is to follow another in the streams of syllables they hear. They can not only figure out where one 'word' begins and another ends, they can also detect the presence of these 'words' hidden inside collections of syllables that are not words. And they can do this even though they've only heard the nonsense syllables briefly. More recent work by Saffran and colleagues has shown that the same results emerge with real languages, as opposed to using made-up streams of syllables, and the ideas have even been extended to music, and to seeing shapes. All of this suggests that children have, and can use, quite powerful statistical abilities to analyse the world around them.

The ability to impose discrete units on continuous streams of sound is not, we have seen, uniquely human. It turns out that the ability to use the statistical organization of syllables is also not uniquely human. Newport and Aslin teamed up with the primatologist Marc Hauser in 2001 to test whether certain kinds of monkeys could also do this.[8] The two groups of researchers worked together to see whether cotton-top tamarin monkeys could also extract the same groups of 'word-like' elements from streams of nonsense syllables. It turns out that they can. You can redo the original Saffran-Aslin-Newport experiment with these monkeys and get a result that is very similar to what was found with human children.

On top of this, there's an intriguing twist to this finding. Research on this topic led by Newport in 2004 showed that what really matters for humans, in these particular experiments, is the consonants or the whole syllables.[9] Humans can extract word-like units from streams of syllables by paying attention to how likely certain consonants are to follow others in the streams of sounds (ignoring the vowels), but they can't do this—or at least not very well—by paying attention to the vowels and ignoring the consonants. The cotton-top tamarin monkeys, in contrast, can pay attention to the vowels, and not the consonants. This means that the ability to see and use the statistical organization of the sounds is common to human children and this particular kind of monkey, but humans and monkeys use different aspects of their experiences.

What this emphasizes is that some of the general abilities that we humans have, that enter intimately into our capacity to use language, are shared with other species of animal, but that the details of how those abilities work may differ from animal to animal.

Backing up this conclusion, Newport's team showed that humans are very bad at figuring out statistical patterns when the patterns skip a syllable, but Tamarin monkeys are pretty good at this. The way they did this was to invent a language with nonsense syllables where there was a high degree of likelihood that a certain syllable, say, *di*, would be followed in the next but one syllable by, say, *ta*, with other random syllables in between. The statistical patterns of likelihood in this kind of made-up language involve syllables that aren't directly next to each other. However, the statistical patterns are very strong—whenever you hear a *di*, the next syllable but one will definitely be a *ta*, and so on. Newport's team found that monkeys were better at this language learning task than the humans—Planet of the Apes here we come.

But it's not just monkeys. Rats are also better at learning certain patterns in syllables than we are.

A team of researchers in Barcelona led by Juan Toro has also been investigating whether the difference between consonants and vowels matters to children learning a language.[10] This team looked at whether children learned linguistic patterns involving consonants better than ones involving vowels, or vice versa. They didn't test likelihood though; they looked at pure and simple linguistic patterns. They showed that children quite easily learned a pattern of nonsense words that all followed the same basic shape: you have some consonant, then a particular vowel (say e), followed by another consonant, that same vowel, yet one more consonant, and finally a different vowel. So words that follow this pattern would be *dabale*, *litino*, *nuduto*, while those that break the pattern are words like *dutone*, *bitado*, and *tulabe*. Toro's team tested eleven-month-old infants, presenting them with words that followed the rule, then, once they'd got used to these, testing out words that either followed, or broke, the rule. They found that the infants learned the pattern pretty well.

However, when the rule was about consonants, the children just didn't learn it. When they were presented with words like *dadeno*, *bobine*, and *lulibo*, which have the same first and second consonant, they just didn't see this as a pattern. The researchers trained children with these words, just as they did with vowels, but when they tested them, the children reacted just the same to *didola*, which follows the rule, and *boneda* which breaks it. Rats, incredibly, when tested with the same experiment, learned both the vowel rule and the consonant rule.

The Newport and Toro experiments are about slightly different things. Newport's team looked at how likely certain consonants are to follow others. Their idea is that those patterns of likelihood allow children to pull words out of the speech stream. They showed that human children were sensitive to patterns of likelihood that involved consonants, but not to those that involved vowels. Tamarins were sensitive to the opposite pattern.

Toro and colleagues looked at whether infants and rats could learn rules about what a possible word could look like. They showed that children could learn a pattern where the consonants were kept the same, and the vowels change, but not the opposite. Rats were sensitive to patterns that involved either consonants or vowels. They didn't discriminate.

What both teams showed is that humans are sensitive to the consonant vowel distinction, and that consonants are a kind of stable anchor. Children attend to the particular consonants, subconsciously expecting them to be broadly stable if they are to form a word. They don't have the same expectations for vowels. These are expected to change while the word stays constant, and they have no special status in defining what a word is.

Monkeys and rats, in contrast, don't share such expectations. They look for patterns in the data of any sort. They aren't limited in what they subconsciously expect to see, and, so they see patterns that are invisible to human babies.

The upshot of these experiments is that human children can keep track of certain kinds of statistical patterns in language that animals cannot (for example, likelihood of certain consonants following others), but animals can do way better than humans at others (for example, likelihood of certain syllables following others when a syllable is skipped). Human children are very good at detecting rule-like patterns in words when the pattern is one that involves vowels, but not when it is a rule that involves consonants. Rats, however, are good at both.

These experiments really drive home that we have certain limits to how we perceive and process linguistic experience. Our brains unconsciously and automatically bring biases to the task of processing what we hear. It's those limits that give us the particular structures of human language.

This makes logical sense too. There are huge numbers of possible aspects of a stream of sound that could be relevant to some kind of pattern or other. But we humans all converge on quite a limited collection of linguistic properties. This is because our brains are set up in particular ways to pay attention to just certain aspects of what we hear. As little babies, we automatically chop up that experience in certain ways, and pay attention to just particular regularities in it. The statistical patterns matter, but they are not the only thing that our minds bring to this task. We've just seen that our minds, unlike those of monkeys and rats, process consonants and vowels in very particular ways.

There's interesting evidence when we look at languages for this difference between consonants and vowels too. The Semitic languages, including Hebrew, Arabic, Amharic, and Tigrinya, have a very special way of organizing their words, built around a system where each word has three consonants, but the vowels change to tell you something about the grammar.

For example, the Modern Hebrew word for 'to guard' is really just the three consonant sounds *sh-m-r*. To say 'I guarded', you put the vowels *a-a* in the middle of the consonants, and add a special suffix, so you get:

shamarti

But to say 'I will guard', you put in completely different vowels, in this case *e-o* and you signify that it's 'I' doing the guarding with a prefixed glottal stop (remember we write this as ʔ):

ʔeshmor

Here, we see the same three consonants *sh-m-r* staying stable, but the vowels are changing, as well as the affixes, to make past versus future Tense.

You can also make this word into a noun, meaning just 'guard'. The Singular form of this, for a male guard, is *shomer*, again, same consonants, different vowels.

We can even see this a bit in a language like English. The present Tense of the verb 'to ring' is just *ring*, as in *I ring the bell every day*. The past is, however, *rang* (*I rang the bell yesterday*) while you use a different form in *The bell has now been rung*. Same consonants (*r-ng*), but different vowels.

Our particularly human ability to store the likelihood of certain consonantal patterns underpins this kind of grammatical system. We can learn grammatical rules that involve changing the vowels quite easily, so we find languages where this happens quite commonly, and some languages, like the Semitic ones, make enormous use of this fact. We do actually also find some languages where consonantal patterns alter for grammatical reasons, so we know such rules are learnable as part of human language. For example, the verb *lu* in Ancient Greek means 'to loosen'. To say 'I have loosened' you take the first consonant, followed by the vowel *e*, and prefix it to the verb, giving *leluka*—*ka* at the end signifies that it is the speaker who has loosened something. This is how the general pattern works in Ancient Greek: *pemp*, meaning 'send', gives *pepemptai*, 'I have sent', and *tlā*, meaning 'suffer', gives *tetlāka*.

These patterns are not used as the whole blueprint for a language. Imagine a language that is like the reverse of Semitic: the words are fundamentally patterns of vowels, and the grammar is done by changing the consonants around the vowels. We could make up a language that worked like this, no problem. It would be pretty impossible for a child to learn, though, if the evidence we've just discussed holds up. We also know of no language that works like this. Consonants anchor words, not vowels. This suggests that the way our particularly human brains are biased to certain kinds of linguistic patterns, but not to other equally possible ones, has had a profound effect on the languages we see across the world.

There are other aspects of the speech stream that seem to be important for our ability to distinguish and learn discrete words. One of these is word stress.

What is a word stress? It's just the idea that each word has a syllable in it that is pronounced with more force than any other syllable. Say the words *organization* and *organize*. In the first of these, the syllable written *za* is pronounced with most force, while in the second, it's the first syllable *or*. Exactly where the syllable with most stress appears is broadly predictable in English, though not always. But many words have a stressed syllable, and those that do not tend to cluster together with those that do.

Elizabeth Johnson and Peter Jusczyk in 2001 showed that eight-month-old babies—the same age babies as Saffran and her co-workers studied—use stresses to pick out words, in preference to information about how likely one particular syllable is to follow another. Later work by Jenny Saffran and Erik Thiessen in 2003 showed that this is an ability that grows as children develop, as seven-month-olds seem to prefer to use likelihood, while twelve-month-olds prefer stress. More recent work in 2014 by the computational linguists Benjamin Börschinger and Mark Johnson has shown that highly sophisticated statistical approaches that go far beyond the simple likelihood of one syllable following another are also improved when stress is taken into account.[11]

It's probably unsurprising by now, but other animals are also sensitive to stress in a speech stream. Zebra finches, budgerigars, and, of course, rats can distinguish words by their stress pattern, though no one has yet shown that these animals are able to recognize where discrete words are in a speech stream by using stress information. So once again, it seems that animals and humans have a similar ability to use aspects of the sounds they hear in the same way, but they use them differently.

There are three important ideas to take away from all of this.

One is that even very young children are desperately seeking discrete words in the language they hear around them: children learning languages impose discrete categories on what they hear at at least two levels. The continuous waves of physical sound are turned into discrete mental sounds. At a higher level, children group these into discrete chunks, which roughly match up to what we intuitively think of as the words of the language.

The second important idea is that children appear to be combining various sources of information to figure out what the discrete words in their language are.[12] We've seen that they can use the statistical likelihood of one sound following another, the idea that there's a single stress per word, the pattern of consonants and vowels in the speech stream. More recent work has shown that they also can use information about the rhythmical patterns in the language, the grammatical endings of words, and the ways that patterns of pauses when we speak match up with the starts and ends of words.

Once again, the question of limitations comes up, and this is the third important idea. There are lots of things children could track in the language they hear around them. They might, for example, focus on the pitch or tone of the syllables. They might pay attention to how the blinking of the eyes of their parents matches up with word boundaries. They might pay attention to the patterns of vowels, as opposed to consonants to figure out linguistic rules. But, at least as far as we understand things at the moment, they don't. They subconsciously limit their attention to certain very specific aspects of their experiences.

I've focussed so far on teasing out how we use patterns of sound to determine the discrete words of our languages. But words are more than just sounds. Words have meanings. We store thousands of words in our heads. These words bundle together meanings with discrete patterns of sounds, or in sign languages, a pattern of shapes of the body. How do we, as children, come to associate the sound with the meaning?

Willard Van Orman Quine was a distinguished philosopher whose career spanned much of the twentieth century. One of his many contributions to how we think about language is a fictional story he told about a linguist out in the wilds, learning a previously unknown language.[13] Quine asks us to imagine that our linguist is sitting down with one of the native speakers of this language, and a rabbit hops by. The linguist points at the rabbit, and utters the phrase he's learned for 'What's that called?'. The native speaker responds *gavagai*.

Those of us who have done field linguistics have this kind of thing happen to us all the time. What I'd do in this situation is get out my field notebook, and write down the word *rabbit* and next to it the word *gavagai*—though I'd write the latter in phonetic script. But, asks Quine, is that justified? How did I know that the speaker meant 'rabbit', and not 'something hopping', or 'floppy ears', or 'a grey thing', or, even, in one of Quine's more memorable phrases 'a collection of undetached rabbit parts'? What Quine is asking is how we can know what our native speaker means, when all we have access to is what she says, and, by assumption, we don't speak the same language.

The common sense answer to this is that we are both, linguist and native speaker, humans, and we both see the world in similar ways. Objects of a certain size, a certain coherence, with certain properties, like being alive, or moving as a unit, matter to us humans, irrespective of our culture. We understand the context of an utterance in the same kinds of ways, and that allows us to be reasonably sure that when I point at a running animal, and the speaker says *gavagai*, and we haven't come across a word for 'rabbit' yet, we can both assume that we've effectively communicated the right meaning. But that's a lot of shared background.

Imagine an alien race, completely different to us, physically, cognitively, and culturally. Perhaps they don't see the world in terms of objects but rather in terms of processes. For the aliens, a rabbit running across a field and a fox running across the field would both be expressed by the same sentence, something like *there's some running going on and it's taking place across a location.* To express the difference between a rabbit and a fox, these aliens would need to refer to different kinds of processes that maintain life, perhaps vegetarian versus carniverous digestion.

For our aliens, who see the world in terms of processes, *gavagai!* would not pick out the rabbit, but rather the event. Now perhaps eventually, with a lot of work, applying techniques and methods that linguists have honed over generations, we'd come to understand enough about how the alien language worked. But say we were not linguists, with lots of training, knowledge, and thinking time. Say we were human infants. How could we learn what *gavagai* means? Unlike the field linguist, children haven't got a lifetime of experience of the world to help them guess what people are talking about when they speak.

The child psychologist Elizabeth Spelke, in her Harvard lab, and her colleague Susan Carey, in hers, have investigated many aspects of how human infants come to know what they do about the world around them. Spelke comes to the conclusion that children are born with what she calls 'Core Knowledge'.[14] This is a basic set of ways that children interpret the world, which, Spelke, argues is common to all humans. For Spelke, Core Knowledge includes how children understand objects in motion around them, how they can be grouped together in countable ways, and the shape of the spaces in which they move. It also includes how children understand the intentions of thinking beings, and social relations between them. These various aspects of Core Knowledge are bound together by their acquiring a language.

What makes Spelke think this? Just like older children and adults, babies can be surprised when the world around them doesn't behave as they expect it will. Spelke has taken advantage of this fact by designing studies where something weird happens in the situation that babies are looking at. She then measures whether the babies or infants have a reaction of surprise. You can do this even with tiny babies by measuring how fast they suck on dummies. Babies tend to suck on dummies at a pretty steady rate, but when something unexpected happens, their sucking speeds up. So you can use special electronic dummies which send signals to a computer which can measure the rate of the sucking. You can then test whether the babies are surprised by what they are being shown, by recording the speed at which the sucking changes.

For example, the babies might be shown a scene in which something that has behaved as a solid object might suddenly act as though it is permeable. For example, a ball might drop down from the ceiling, and bounce a number of times, but then a toy car might drive right through it. In another situation, a figure of a person might just disappear, magically, as though it was teleporting. In yet another, one ball might roll towards another, then the other might move, as though it's being affected by the first, even though they are not touching. In all of these cases, little babies behave differently than if the scene before them unfolds without violating their expectations. Children don't expect the world around them to involve magic, and they are surprised, and often delighted, when it does.

Spelke and her co-workers have shown that even babies who are no more than days old have quite firm beliefs about how the world should behave: when what they see moves as a connected whole, and maintains its shape and size as it moves, and stays together as a single object, babies treat it as a single discrete object. For example, babies can be shown an object moving behind a screen, then, the screen can be removed. If the object is just where you'd expect it to be, the babies are interested, but not very much so. But if the object has disappeared, or become two objects, the babies act surprised, and Spelke's team have measured just when that surprise occurs.

From these kinds of experiments, it's become clear that babies treat, as discrete objects, things whose parts move together and cohere. So cups, or trains, or, indeed, rabbits, are treated as objects, but at least initially, babies don't react to piles of sand, or handles of cups, in the same way. Over many years of refining these experiments, it has emerged that the cohesion of an object across space and time is what, for infants, makes something an object. Interestingly, this history of cohesion is far more important than the way that the object looks—once again we see the importance of discreteness in the way that humans think. Once a child has decided something is an object, that thing persists as an object in the memory of the child.

Spelke's work shows that human babies are born with systematic ways of imposing structure on the world they experience.

They experience the world as humans, and not as frogs, lions, or aliens. We could imagine an alien Spelke on an alien world doing the same kind of experiments, and discovering that alien babies see the world as processes, not objects.

Human children perceive the world around them as involving discrete objects; we've seen that they are quite good at extracting patterns from the continuous stream of sounds around them, and deciding that those patterns are words. Is that enough to solve Quine's *gavagai* puzzle? Not quite.

Imagine you are not a field linguist, but a child. You're playing in the garden, which is full of interesting things. The sun is shining, birds are singing, there are frogs croaking from a pool into which a fountain is pouring, and a rabbit is hopping by. Oh, and your mother is pointing, and saying *gavagai*. You automatically see the rabbit as a coherent object, and you've used your statistical skills, and figured out where the stress goes, so you take *gavagai* to be a word. If your mother helpfully points at the rabbit while saying the word, and you know that pointing is somehow something which connects the sound made by your mother speaking to the object being pointed at, then you're more or less there. That's the third part of an explanation for how to solve the *gavagai* problem: speaking and pointing is a mechanism that is inbuilt in you that allows you to connect sounds with things. As a child, you are able to understand the intention of your mother in making the noise and pointing to an object. Again, this is part of being human, though not unique to being human—other animals have been found to understand that pointing can be used to direct someone's attention to something, though I must admit, my cat is frustratingly useless at it.

This is a very simple and pretty common sense type of theory of how children learn words. But it's already quite complex. It needs children to be mentally set up to understand the world as consisting of discrete objects, to be able to extract discrete units out of continuous streams of sounds, to use a rule of thumb connecting stressed syllables to words, and to understand that pointing at an object and speaking at the same time is to be understood as highlighting a symbolic connection between the sound and the thing. So the child, at a very early age, has to have a bunch of fairly complex and specific abilities. The child also has to coordinate these abilities in quite a focussed way so that they operate to help her to learn the words of the language. And something quite miraculous, when you think about it, is that all children in all cultures do this within the first few years of their lives.

But this common sense explanation is still nowhere near enough. Children don't just learn nouns, like 'rabbit'. They also learn verbs, like 'hop', and they learn verbs even when the event the verb describes isn't happening in the situation around the baby. Large amounts of speech to children don't involve a description of what is happening. Researchers have found that in more than a third of full sentences spoken to children, the nouns don't pick out things in the surrounding situation. Somehow the children must deal with many cases where nouns are being used to talk about things that are absent. On top of this, sentences might be in the past, or the future, or they might be about what could happen, or should happen, or isn't happening. They might be about feelings, or beliefs, or other abstract things that don't have concrete shapes or motions. So children have to go well beyond just the simple idea that saying something and pointing to it is enough to learn the words of a language. They also need to keep track of the connection between a sound and its possible meanings across many many different utterances.

If you own a dog, you might not be that surprised that babies can connect words to their meanings in this way. A paper published in *Science* in 2004, introduced Rico, a Border collie, who showed a pretty impressive ability to connect sounds and things.[15] Rico was trained first by his owners, then by the researchers, and he ended up with a capacity to link more than 200 objects to sounds. The way that Rico did this is similar to an idea about how children learn words. If you see a bunch of things, one of which you don't know the word for, then you hear a new word, you associate it with the thing you don't know.

But though Rico is a very impressive dog, he required quite specific training to learn the words for new objects, while infants and toddlers pick up much of their vocabulary without explicit instruction. And Rico is a pretty special Border collie, which is why he had a paper published about him in a top scientific journal. In contrast, all children learn thousands upon thousands of words with no special training. A normal dog will learn maybe five or six words (sit! stay! lie! fetch! down!). Working sheepdogs learn many more. Rico learned 40 times what the average dog learned. Just think how odd it would be to single out one specific child as being able to learn 40 times more words than an average baby. We just don't see this with humans. Human children differ in how quickly they learn words, certainly, but the variation in this is just normal variation. There is a great story—untrue, unfortunately—about Thomas Macaulay, a famous nineteenth-century British politician and historian. Apparently, when he was a tiny baby, he was taken out to have afternoon tea by his mother in his baby cot. His hostess accidentally spilled a hot drink on him. Like all babies, he screamed his head off and his hostess fussed over him, until he reportedly said, 'Thank you Madam, the agony is sensibly abated.' Macaulay would be like the Rico of children.

However, although Rico needed to be trained by humans to learn his 200 words, there are animals in the wild that naturally use specific patterns of sounds for particular meanings. We see this especially with various kinds of monkeys, perhaps most famously with vervet monkeys. Bob Seyfarth and his co-workers, Dorothy Cheney and Peter Marler, working in the African savanna, tested observations that had been made about the calls of these monkeys back in the 1960s.[16] The researchers in the 1960s had seen the monkeys making different kinds of calls depending on whether there were different kinds of predator around. One particular call for leopards, one for snakes, and another for eagles. Seyfarth's team tested whether the calls were really used to communicate, as opposed to just being reactions to what was going on around the monkeys (like yelling 'Owch!' when you are hurt). Seyfarth and his team recorded the monkeys making their calls, then played back the calls to the monkeys to see what they'd do. The monkeys reacted to the different recorded calls just as though the relevant predator was around. If the team played the call that monkeys would usually make if there were an eagle circling, the monkeys would dive from the trees into brush cover. If he played the call associated with leopards, the monkeys would run into the trees. It seems a little harsh to go around terrorizing monkeys, but we learned from this that the calls could be understood by other monkeys as carrying information about the environment. The calls are not just reactions to the situation, they are treated as though they are warnings.

It turns out that many species of monkey have distinct calls for distinct predators. Not just vervets, but also Diana monkeys, putty-nosed monkeys, tamarins, ring-tailed lemurs, and the Campbell monkeys, who live in the Tai Forest in Côte d'Ivoire. For some of these monkeys, what is crucial is whether the threat is in the air or on the ground, while for others, it seems to be more specific to the actual predator. As well as calls being related to predators, certain monkeys and some apes seem to connect calls to different types of food.[17]

But there are big differences between monkey calls and the words of human languages. Human words outnumber monkey calls by a factor of tens of thousands. Monkey calls are, in terms of their sound structure, massively simpler than human words. Although monkeys can, and do, alter how their calls sound, the basic structure of the call seems to be genetically given, so that monkeys raised by other species keep to their own species calls. But perhaps the most striking difference is that, as far as we know, monkey calls are what linguists call 'referential'. There is a direct link between the call and the thing in the world that the call is connected with. Human words seem to be fundamentally different from this.

Let's think about the word 'thing' for a bit. Imagine we are walking around a modern art gallery together, and we see a collection of bricks haphazardly strewn across the floor of a room. Are these bricks a thing? Or, more accurately, can we say that the word 'thing' can be associated with the bricks? Well, it depends on whether the bricks are just the left over debris of some building works, in which case they are things, but we wouldn't want to say that they are a 'thing'. But what if they are a carefully placed work of art, that's just won a big prize. In that case, it would be fine to say 'What did you think of that thing in the second floor room? Should it have won the big prize, do you think?'. How we use even so basic a word as 'thing' depends on our perspective for its meaning.

Or, let's take the word 'pawn'. Imagine I have 16 brussels sprouts on the kitchen table. Are they pawns? Definitely not. But now let me array them on a chessboard, so that each one is on a square on the second row. And now I'll put a king, a queen, two bishops, two knights, and two rooks behind them. Now, let's play chess. Can I say to you, as I move a brussels sprout up two squares, 'I've moved my pawn to King four'? Of course! Even though a brussels sprout is not a pawn, I can use the word 'pawn' to talk about a brussels sprout, as long as I'm thinking of the sprout as having the properties that I'd assume a pawn has. Words don't only depend on the perspectives of the speaker, they depend on the intentions of the speaker too. The calls of animals, as far as we can tell, don't work like this.

Words can also be used to talk about things that don't exist. Harry Potter flew on a hippogriff at Hogwarts. Who flew on what where? Hundreds of thousands of people would completely agree with that sentence, and be quite happy even to say that it is true. But there's no link between the words *Harry Potter* and an object in the world that flew on a hippogriff. If there isn't anything in the real world to associate with a word, the uses of words, and their meanings, must go beyond just connections with things in the world. In this case, everyone who knows what that sentence means shares a fictional world, a world of the imagination, a world constructed by J. K. Rowling that many people have internalized. We *create* meaning by using the words of our languages.

This is different from the monkey calls. These can't be used by the monkeys to mean what they want them to mean. For the monkeys, there appears to be a direct link between the call and a thing, which is why they, quite sensibly, scurry up trees when they hear the call associated with leopards. But human words, as we saw with David, the homesigner, are the very stuff of creation. We humans impose discrete categories on the continuous sound we hear. We also impose structure on our perceptions of the world when it comes to concepts. A word is the connection between these, learned from our exposure to the language around us as we grow up. Unlike other species, the connections between sound and concept isn't baked into our genes; we learn these connections as the basic discrete units of language. But though the connections themselves are learned, what gets connected, the sound and the meaning, is structured, organized, and shaped by the limits of the human mind.

The capacity to conceive of the world as involving discrete units, even though the gross physics is continuous, is a capacity we share with other animals. The capacity to connect discrete patterns of sounds to concepts is also something we share with other species, and Darwin was correct in saying that we are distinguished from other animals in terms of sheer brain-power: we can do this connection over tens of thousands of such patterns, whereas other animals are severely limited to a few tens of calls.

However, the concepts we connect with words are different in kind from the concepts that, say, monkeys associate with their calls. Ours can be non-referential and, in fact, are almost entirely so, but animal calls are directly connected with their environment.

The grammatical categories we met in the last chapter, things like Person, Tense, and Number, are prevalent in human languages, but don't exist in the animal world. Monkeys don't have a word for 'I', they can't talk about what just happened, or what might happen, they can't say 'Here come three leopards'. Human words are not souped up animal calls.

But perhaps the most striking difference between animal calls and human words is that the latter are organized syntactically in sentences. As we've seen, this syntax is hierarchical, and abstract, and links to meaning in systematic ways. There's no evidence that other species have this ability with words. Is this just a side effect of human's more expansive capacity to learn? Do we *learn* the hidden hierarchies of our languages?

7

A LAW OF LANGUAGE

Kanzi is a bonobo, a member of an endangered species of chimpanzee from the Democratic Republic of the Congo. He was born to a bonobo called Lorel in 1980 in the Yerkes Primate Center in Atlanta, Georgia, but swiftly adopted by another female called Matata. Matata herself was the subject of a research project to teach apes a kind of simple symbolic communication system. The system took advantage of the idea that apes have rich conceptual abilities, and can associate discrete symbols with concepts. It used pictures on a board for this purpose. The researchers called these pictures lexigrams. The idea behind the lexigrams was to bypass the fact that bonobos' very different vocal anatomy makes imitation of human speech sounds impossible for them.

Matata, to the researchers' chagrin, was somewhat uninterested in this whole experience. Her adopted child Kanzi, however, was fascinated by the lexigrams and spontaneously began to use the board to communicate with the researchers. Most of Kanzi's communications were functional: he wanted food, or toys, or to go somewhere. Very few were descriptive, telling the researchers about how he saw the world. Nevertheless, his ability to communicate was quite impressive.

Susan Savage-Rumbaugh, who led this research, has argued that Kanzi's abilities with language are comparable in some ways to that of humans.[1] Kanzi's communications using lexigrams were quite limited. They tended to involve a lot of repetition, and there was no unequivocal evidence of linguistic structure in them. But, Savage-Rumbaugh argued, Kanzi also had an ability to understand English sentences, an ability that, according to Savage-Rumbaugh, was comparable to that of a young human child.

To demonstrate this, when Kanzi was about nine years old, Savage-Rumbaugh carried out a series of experiments. She placed another experimenter, together with Kanzi, behind a one-way mirror—this was so that Kanzi couldn't pick up unconscious cues from Savage-Rumbaugh herself. Savage-Rumbaugh then spoke to Kanzi, asking him to do various tasks with the other experimenter. Kanzi's various actions were recorded, and graded for how correct they were. Savage-Rumbaugh gives the overall level of correctness at about 70%. So they could have a comparison, the same test was carried out with a young child of about two years old, Alia, who scored just a little lower than Kanzi (about two thirds correct).

The linguist Steven Anderson has suggested that Kanzi's performance is the result of what he calls a 'semantic soup' strategy. The idea is that apes are clever creatures. They can associate words with concepts, just like Rico the Border collie, and given the concepts, they know what the most sensible thing to do would be. Rather than grasping a syntactic structure, Kanzi, according to this interpretation, understands various words, and does what is most sensible according to that understanding.

It's as though we humans were to hear an unstructured sequence of words like *beetle chimp eat*, then were to be asked what happened. It's pretty likely we'd come up with the idea that the chimp ate the beetle. But, if I said to you 'The beetle ate the chimp' then you know it was the other way around. The first strategy works off what is sensible. The second uses syntax, and syntax doesn't care about what is sensible!

Is Kanzi using the semantic soup strategy? Or is he understanding the syntax of the English sentences?

Some of Savage-Rumbaugh's instructions to Kanzi were complex, and unusual, but still resulted in Kanzi doing the right thing. For example, Savage-Rumbaugh asked Kanzi to both 'put the tomato in the oil' and to 'put the oil in the tomato', and Kanzi carried out appropriate actions. This suggests that Kanzi can connect the order of the words to the overall meaning in a relevant way. For Savage-Rumbaugh, this is evidence that Kanzi had a kind of ability with grammatical structure that is at least somewhat similar to that of human beings.

Is Savage-Rumbaugh's experiment evidence that Kanzi has a sense of linguistic structure like we do? Rob Truswell, a linguist at the University of Edinburgh, has done a careful study of the way that Kanzi responded to Savage-Rumbaugh in her experiment, and thinks not.[2] Although Kanzi might be able to connect order to meaning, it seems that he's not able to chunk words together into groups, giving the kind of hierarchical structure that humans unconsciously attribute to the sentences of our languages.

To show this, Truswell examined how Kanzi reacted to sentences where the words for objects are chunked into a single piece of structure with the word *and*. This kind of chunking grammatically conjoins the various words into a larger unit, creating a hierarchical structure for the sentence.

Give the water and the doggie to Rose.

Give me the milk and the lighter.

Now, as we would expect, Alia, the human child, treated *the water and the doggie* and phrases like it as a single chunk, acting on both items as a single unit. Alia's behaviour with this kind of example was at the same level of correctness as her behaviour overall (around 66%). This shows that she has chunked the two phrases into a single unit, so she's imposing a hierarchical structure on the sentence, conjoining the words that express what is being given to Rose. We can express this hierarchical structure by a tree diagram, like this:

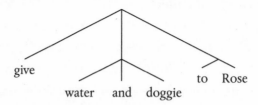

Kanzi, on the other hand, is much worse at these coordinated sentences than he is overall. In fact, Truswell shows that he gets only about 22% of them right, compared to his overall score of about 70%. What Kanzi does with these sentences is randomly ignore the first element, or the second, or take both. Truswell points out that this means he's not sensing the hierarchy at all. He doesn't understand that he's to conjoin the words for the two objects into a single unit, unlike Alia.

Kanzi is using order and not structure. We could represent what he is doing not as a tree diagram, but as a sequence of words, like this:

give—water—Rose

give—water—doggie—Rose

The first of these is easy for a bonobo of Kanzi's intelligence and training to work out. The second leads to confusion. Alia, on the other hand, knows that there's a hierarchy here, and is able to conjoin *water* and *doggie* into a unit.

Kanzi does have an impressive ability to connect not just signs to meanings, but also the order of signs to meanings. He understands the notion of a sequence, and can relate the words in a sequence to a more abstract idea of an action involving various people and things. But he doesn't do what we humans do. He doesn't treat the sentence as though it has a hierarchical organization. He doesn't have the sense of linguistic structure that human children can't help but employ.

∞

Why do humans, but not apes like Kanzi, treat sentences as having an invisible hierarchical structure?

In the first chapter I laid out two different ideas about this. One idea is that we learn the hierarchy. The other is that we impose it through our sense of structure. I'm convinced that the second approach is right, but we haven't yet looked at how the first idea might work in practice.

How would we learn a hierarchy?

We don't get explicit instruction when we are children learning our native languages. Researchers who have looked at how children learn languages agree that explicit correction is rare and that children usually ignore it. The consensus is that children learn the details of their native language by observing what people say around them. Syntactic hierarchy isn't an audible part of a sentence. Occasionally, pauses, or changes in intonation, can provide some information about syntactic hierarchy, but again, the consensus is that that is also insufficient. So where would information about it come from?

An intuitive approach would be to say that we hear a lot of specific phrases and sentences as we are growing up. We'd learn each of these. But we'd also subconsciously notice recurring patterns in those phrases and sentences. Humans have very general skills that let us extract patterns from the world around us and extend these to new experiences. Using these skills, we would abstract away from the specifics of the phrases and sentences we've heard to get at more general patterns. Once we have the general patterns, we can use these as a kind of model to make up new sentences. This would be a way of essentially learning our sense of linguistic structure.

This set of ideas is the major alternative to Universal Grammar as a means of explaining how human language syntax works. It has been developed by a number of researchers across the world. Because it proposes that children learn specific grammatical constructions first, researchers who advocate it go by the name of constructionists. Constructionists argue that there is no Universal Grammar and that all the properties of syntax can be learned by children from their experiences using general intellectual skills.

But what specifically are these general skills? The linguist Joan Bybee, a leading constructionist, has suggested that at least two are crucial in understanding how we create new sentences: Chunking and Analogy.[3] Bybee's suggestions have been extended and developed by other researchers, and Chunking and Analogy have become the major tools of explanation in constructionist approaches to language.

Constructionists and advocates of Universal Grammar generally agree that the patterns in languages are, indeed, best captured by hierarchical structure. Constructionists say that we humans can use our powerful general intellectual skills to discover that structure, while advocates of Universal Grammar propose that our minds are tuned to language, and we impose certain kinds of structure on it.

For Bybee, an ape like Kanzi would simply lack the intellectual power to discover linguistic structure. His Chunking and Analogy abilities are just weaker than ours. Such an approach is consistent with Darwin's idea that humans are distinguished from animals by our general intellectual powers, rather than because of a particular set-up of our minds.

Let's look more closely at Chunking and Analogy and see whether these general skills could explain how human language syntax works.

Repetition of units that are heard together a lot triggers our minds to make what Bybee calls a 'chunk'. A chunk is just a piece of memory which consists of other things which are already in your memory separately. So if you remember your bedroom, and you remember your cat, and you see your cat in your bedroom and remember that, that last memory is a potential chunk. If you see your cat in your bedroom again, that experience makes the chunked memory more solid. More experiences of similar situations build stronger and stronger associations between cat and bedroom.

I'm always amazed when, on hearing a song that I've not listened to for years, I can still remember lyrics. This is because that music and those lyrics have been chunked in my memory many times, and are therefore stored way deep down in my mind. A more linguistic example would be the particular meaning of an idiom, like *drive me crazy*. You hear this phrase in particular situations where something has made someone irritated, and that connection, after time, gives you the idiomatic meaning of the words.

Chunks get stronger the more you experience them. No repetition, no Chunking. Bybee argues that Chunking is what gives the grammar of language its particular shape, with larger sentences made up of smaller words. Each word is a chunk of sounds and meaning. When you hear particular words sequenced together often enough in certain situations, the sequence becomes a larger chunk associated with the reoccurring situation.

Sometimes the bits of the chunks will vary. Rather than *drives me crazy*, you might hear *drives me mad*, *drives Lilly mad*, and *drives Lilly insane*, and what you end up storing is a chunked pattern with abstract slots in it, a bit like this:

SOMEONE DRIVE SOMEONE-ELSE X
means *someone makes someone else feel some kind of emotion*

The kind of things that X can be depends on what you've heard in that X slot. In this case, they are words or phrases that mean something like *crazy*. In fact, the words *crazy* and *mad* are very frequent in that slot, so they serve as a kind of model for what can go there.

A second kind of general skill that Bybee suggests humans use in language, together with Chunking, is Analogy. Analogy is a process which allows speakers to use a new item in a chunk in a way that they have never experienced themselves. For example, you've probably heard the chunked pattern above used with the word *bananas*, as in *He drove us bananas!* At some point, someone was the first person to use this. There was a novel extension of the pattern, perhaps aided by the use of the word *bananas* to mean crazy (as in *that idea's just bananas!*). This is Analogy. You extend the chunked pattern in a new way. So putting Chunking (with slots) together with Analogy, you begin to have an explanation for why people can use sentences in new and creative ways.

But not just anything goes. Bybee points out that it would be weird to say *That drives me happy*, since, while the slang meaning of *bananas* is similar to that of *mad/crazy*, the word *happy* is quite different. So Analogy is constrained by similarity.

Bybee argues that both Chunking and Analogy are general facts about how we think; they are not tied to language, but humans use them when they are exposed to language. For example, think of some socks. Now think of some feet. There's a relation between these. Now think of hands. If we keep the relation between feet and socks in mind, then hands bear that same relation to gloves. That's Analogy.

You've come to know my cat Lilly so far. When she makes a sound it's a miaow. I don't have a dog, but if I did, the analogous sound would be a bark. That's Analogy.

Analogy is not specific to language, but Bybee and others argue it is used in language to create rules. For example, perhaps you hear the verb *kick* then you hear *kicked* in the past Tense. Once you've heard enough examples like that, you use Analogy to work out that, as *kick* is to *kicked*, so *kiss* is to ... *kissed*. Analogy allows exceptions, so you can learn those independently when you hear enough examples. Although, as a child, you might come across *go*, you never hear *goed* for the past. You always hear *went*, which allows you to store an exception to your Analogy.

If we have Chunking, and we have Analogy, then, we might have a way of extracting syntactic patterns from what we hear as we grow up. The chunks would provide us with a pattern, and Analogy would let us extend and generalize it, creating rules.

If these ideas can be made to work—and I'll argue that they can't—then we could say there's nothing special about human language. Languages are just the outcome of how these skills of Chunking and Analogy, and a few others, have affected how people speak over many generations. From Bybee's perspective, there is no Universal Grammar. There's nothing about language that makes it fundamentally distinct from other aspects of human culture, like dance, or art, or mathematics, or cooking.

The alternative answer to the question of how our abilities with language allow it to be used creatively is this: our minds are set up with a system of unconscious Laws of Language (Universal Grammar). These connect, and constrain, what we mean and what we say. We use the raw material of our experiences, together with these laws, to create a system of rules for how our own particular language works. The rules are what we use to create new sentences as we need them. They are the sense of structure I talked about earlier in this book. The reason that Kanzi failed with conjoined phrases while Alia didn't is because Alia's mind is set up in a different way. She has a sense of structure that makes her interpret the sentences hierarchically, while Kanzi lacks this. Alia, at least at her age, doesn't have a more powerful intellect than Kanzi, her mind is just set up differently for language.

In previous chapters, I've discussed some of the evidence that makes me convinced that we humans interact with aspects of what goes on around us by turning it into a particularly human linguistic system. The core of language is in us, and the particular shape that core takes as it develops depends on a to-and-fro between our nature as linguistic beings and the language(s) of our communities. One kind of evidence came from home-sign: homesigners have no experience of repeated chunks, as the gestures that they are surrounded by don't have linguistic structure. Yet homesigners create grammatical structure in their signing. Another argument comes from how homesigners create full languages so quickly when they are brought together, as in Nicaraguan Sign Language, or the Zinacantán signers. On the constructionist view, languages are shaped by slowly gathering structure, through Chunking and Analogy, as they are transmitted down the generations. But that leaves us with no explanation for the speed with which quite complicated structures emerge in some new sign languages. Yet another argument comes from the absence, in languages of the world, of conceivable and feasible types of languages that, nonetheless, do not exist. Brain scan evidence gave us another argument that human beings are set up with a sense of linguistic structure, distinct from general skills in thinking. The experiments showed that people use different parts of their brains to process possible versus impossible languages, suggesting that there are particular types of brain activity connected to language, as opposed to other kinds of mental work.

In what follows, I want to give a different kind of argument against the idea that language is a result of Chunking and Analogy. I'll make the argument from the way that patterns of language work in one particular language, English. I'll show that Chunking and Analogy can't explain these patterns: Analogy makes the wrong predictions. Instead, what is needed is abstract structures. I'll argue that these structures are restricted by general laws. Just like in physics or chemistry, these general Laws of Language can give us an understanding of why particular languages, in this case English, work the way they do. I'll also show you that Chunking is too limited to give us the richness of human language. Even together, Chunking and Analogy are not enough.

We're going to look at just a very tiny part of the system of one particular language, but the conclusions I'll make are general. We could choose to look at almost any phenomenon in any language and, if we look deep enough, the same issue will come up: Chunking and Analogy aren't enough. We can't explain language without linguistic principles, or Laws, that organize our knowledge of it. These principles are unconscious but real.

∞

The particular bit of English we're about to plunge into is how pronouns (words like *she* and *he*) and names (words like *Anson* and *Lilly*) work together in sentences: how we use these words to keep track of the people or animals or things we are talking about. When we begin to look carefully at how pronouns work, there's no getting away from abstract structure.

Imagine we are at a race. My partner, Anson, has been running races of various lengths for a while, but hasn't won one yet. He's just finished and we're all waiting to find the results. I go and get a coffee, and when I come back, I see Anson standing by the results board, his face glowing with happiness. I turn to you and ask, 'What happened?', and you reply:

Anson just learned that he won the race!

By using this sentence, you obviously want to express the thought that Anson just learned that Anson won the race and that's why he looks so happy. But you don't usually use the word *Anson* twice. Instead you introduce him with the word *Anson* then you pick him out again later with the word *he*. You've used both words to pick out the same person, first using a name *Anson* then using a pronoun *he*. You could also have used other phrases, like *your partner*, or *Professor Mackay* in place of *Anson*.

There are some surprising restrictions on how you can use names and pronouns to refer to a person. Imagine you'd said this instead:

He just learned that Anson won the race!

If you'd said this I'd be confused. Who just learned that? You can't use *he* in this sentence to refer to Anson then use the word *Anson* to do the same thing. I'd have to interpret you as meaning that someone else just learned that my partner had won the race, so I'd be justifiably confused.

This is obvious to any speaker of English, but we might ask why it works like this? Sometimes the obvious is quite perplexing. Why can't you use *he* to pick out Anson? After all, you could have pointed at Anson and said:

He just learned that he won the race!

If you'd done that, I wouldn't have been confused at all.

The rule seems to be that you can pick out a person with a pronoun, even at the start of a sentence. But once you've done that, you can't then use a name like *Anson*, or *Professor Mackay* to pick out that same person afterwards. Even if you keep pointing. So imagine just how weird it would be if you'd pointed at Anson, then said:

He just learned that Anson won the race!

Now it seems like someone else must be Anson. This shows us that what is going wrong doesn't have to do with not knowing who is being referred to. The problem is with the sentence, not with what you are trying to convey.

This is how English speakers use pronouns and names. But why does it work this way? What stops you pointing at Anson and saying *He just learned that Anson won the race?*

One answer that seems fairly plausible is that you have used the pronoun before the name. Maybe there's a general pattern, that you've learned over the years, that says something like this: don't use a pronoun to refer to someone before you use a name to refer to them. It'd be a convention that you pick up by watching other people's habits in how they talk. You'd learn this rule for certain sentences, then you'd use the idea of Analogy to extend it to others, just like you extend putting *-ed* on *kick* to make it Past to other verbs. Though this idea seems plausible, it turns out not to work.

Imagine you and I both see Anson coming in first in the race, but he's too exhausted to notice immediately how well he's done. I go off for coffee, or maybe beer for us all, and on coming back, you nudge me and say:

That he's won has completely surprised Anson.

You've just used a pronoun, *he*, to refer to Anson then later in the sentence you've picked him out again using the name *Anson*. If there were a convention barring using a pronoun then a name to pick out the same person, it should stop you saying this. But it didn't. There's no problem, in this sentence at least, using a pronoun then a name to talk about the same individual.

If you think of this in terms of Analogy, it is completely puzzling. Let me repeat the two crucial sentences:

He just learned that Anson won the race.

That he's won has completely surprised Anson.

In the first, we can't use *he* and *Anson* to refer to the same person. In the second we can. They are similar in having the pronoun occur before the name, but different in how the two words can be used to pick out someone in the actual situation.

If Analogy is what is at play, this is not what we expect. Why wouldn't a child use Analogy to get from one of these sentences to a general rule (like they do with *kick* and *kicked*)? And what is the explanation for this difference?

Let's explore an alternative.

∞

The restrictions on when you can use pronouns and names to pick out the same person turn out to depend on the structure of the sentence. In previous chapters we've used diagrams to show the structure of sentences, like chemists' diagrams of molecules show the structure of the molecule. This book isn't about how to do linguistics, it's about what the nature of language is, so I'm only going to scratch the surface of why the structures look as they do. I'll give you a little motivation, so that the idea is clear, and I'll only use structures that are pretty much agreed upon by almost all linguists, so there isn't much that's controversial in what I'll present.

What kind of evidence do linguists use to figure out the structure of sentences?[4] One very simple kind of evidence is when a bunch of words can be substituted by a single word, and the meaning is broadly preserved—we saw this when we discussed David's use of homesigns in Chapter 4. For example, the whole phrase *that he won* can be substituted by a single word *that*, with the meaning staying substantially the same. This works in all the sentences we've seen so far: the word *that* can always substitute in for the whole phrase *that he won*:

Anson just learned that he won   *Anson just learned **that***

That he won surprised Anson   ***That** surprised Anson*

This pattern can be captured by assuming a structure with a slot into which these various units can go. We can diagram it like this:

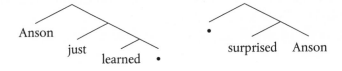

The bullet point here marks a position in the structure that is waiting for something to be bonded to it. You can bond a single word like *that* to it, or a larger structure like *that he won*.

Particular positions have particular kinds of bond. You can see this by contrasting the following two cases, where I've replaced *that he won* with just the word *winning*:

Anson just learned winning.

Winning surprised Anson.

The first sentence is a bit odd. It means something like 'Anson just learned the abstract concept of winning'. The second sentence means more or less the same as *That he won surprised Anson*. The contrast between the two gets sharper if we add *the race* to *winning*:

Anson just learned winning the race.

Winning the race surprised Anson.

The first sentence is now really quite weird, while the second is fine. Different verbs set up different kinds of bond with their Subjects and Objects.

Now let's go back to our original puzzle. Why is it that sometimes you can use a pronoun followed by a name or a descriptive phrase to talk about the same person, and sometimes you can't?

The answer to this puzzle is that English speakers' sense of structure limits how we express what we want to say. There is a set of laws that govern human languages in general, and the particular rules of English have to obey these Laws of Language.

First, let's look at the structure of *He just learned that Anson won (the race)*. We've seen already that *that Anson won* is a single complex unit that can bond with the word *learn*. The structure of the sentence can be given in a diagram like this. I've missed out *the race* so the diagram stays a manageable size:

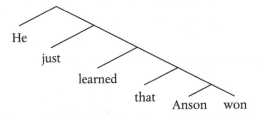

If you put your finger on *Anson* in this structure then just follow the lines up the diagram, pretty soon you come to a line that will take you immediately down to *he*. There is a close structural connection between *Anson* and *he* in this diagram.

Now compare this to the structure for *That he won surprised Anson*, where you can use the pronoun and the name to refer to the same person:

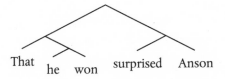

If you perform the same action, and put your finger on the word *Anson*, and trace it up the diagram, to get to the pronoun *he*, you need to go down not just a single line, but right inside a whole subpart of the diagram. To get from *Anson* to *he* you can't just go up, then one down. There's a different, more distant, structural connection between *Anson* and *he* in the second diagram.

The trick I've just shown you, where you trace a line up from the name to a pronoun, is a way of making concrete, an abstract, structural, relationship. It's this abstract relationship that is crucial for understanding what is going on when we use pronouns and names to refer to the same person.

In Chapter 3, I made an analogy between gestures and the mental structures that linguists use these diagrams to talk about. If you think about a gesture made with your hand, you can touch your thumb to the pinky on the same hand quite easily, but you can't touch your thumb to the back of that hand. There's a structural constraint on what connections can be made. Similarly, you can connect *Anson* and *he* in one sentence but not in the other. When the connection can be made, the name can't refer to the same person as the pronoun. A close connection in structure stops there being a connection in meaning.

Here is one way to state this. We can call it the Pronoun-Name Law, and the version I give here is based on the work of the linguist Tanya Reinhart. It forms part of a richer set of Laws discovered by linguists that govern how pronouns and nouns can be used to refer. These are collectively called the Binding Theory:[5]

The Pronoun-Name Law
A name cannot be used to refer to the same individual as a pronoun in a sentence if it is structurally connected to it.

The structural connection can be worked out by the trick of tracing your finger on a tree diagram. But that's just a trick. Tree diagrams just represent the abstract structure of the sentences, the structure we humans unconsciously impose on the sentences. There are real structural connections in the system of language in the minds of English speakers. Linguists have various fancy terms for these, but those terms aren't important for what we are discussing here. What matters is the basic idea: structure, not order, is what is necessary to understand how pronouns and names can be used to refer to things.

This Pronoun-Name Law works, without any real exceptions, for every case of a pronoun and a name in English. Whenever there is a connection of the 'keep going up then go one down' sort, the pronoun can't be used to refer to the same individual as the name. It actually works not just for names, but also for some other kinds of phrases with nouns in them. If we swap in *the professor of paleolimnology* or some other descriptive phrase for *Anson*, we get the same result:

He just learned that the professor of paleolimnology won.

Once again, it's impossible to use *he* and *the professor of paleolimnology* to refer to the same person. This law works extremely well and, as we'll see directly, interacts with other rules of English to give the right results.

Just like a chemist might use the structural properties of a molecule to explain the molecule's behaviour, linguists use the structural properties of these sentences to explain what different kinds of meanings they can express. The Law explains your linguistic behaviour: you can't point at Anson and say *He just learned that Anson won the race* because the Pronoun-Name Law forbids the words *he* and *Anson* being used to refer to the same individual in that structure.

The Pronoun-Name Law works for lots of cases we haven't discussed. In a sentence like the following one, the pronoun *her* is down inside the Subject, *Anson's photo of her*, which I've put in bold:

Anson's photo of her *impressed Karen.*

If you apply the Law to the tree diagram, we can see that *Karen* and *her* are not structurally connected. If you put your finger on *Karen* there's no way to get to *her* by going up, then one down. The Pronoun-Name Law therefore allows them to be used to refer to the same person.

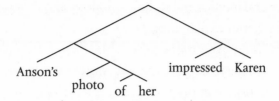

This law also gets our very first sentence right. That sentence has *Anson* higher up in the structure than *he*.

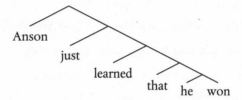

If you put your finger on *Anson* and head up the diagram, you stop pretty quickly, because there's nowhere to go. It's just trivially true that you can't trace a path from *Anson* to *he* with just one downwards step. Because of this, you can use both the name and the pronoun to pick out the same individual, or different individuals.

The Pronoun-Name Law is very different from Analogy, and it is more successful in explaining the patterns of structure and meaning that we see in languages of the world. Once the structure is in place, the Law locks out particular meanings that would otherwise be possible. Its job is to relate structure to meaning. The deeper question of where this Law comes from, and why it works the way it does, are beyond this book.

Is this a better approach to the phenomenon than Analogy? The sentences where there is a pronoun then a name are similar in order, so by Analogy they should behave similarly in other respects. Analogy on its own won't explain these patterns. This isn't the place for detailed linguistic argument, but even if we add in other general intellectual skills to Analogy, the Pronoun-Name Law is still a better explanation.

But there is an even more serious problem for Analogy than just not being able to explain the phenomenon. The big question for a constructionist approach is why children *don't* use Analogy to organize the relationship between pronouns and names? How do they know *not* to use it? Why don't languages change over time so that they all end up using the order of words, as opposed to inaudible hierarchical structure? They don't though. This is what ultimately convinces me that the constructionist approach doesn't provide an account of how the syntax of human languages works.

The Pronoun-Name Law is one of a number of very general Laws of Language, part of Universal Grammar. However, there are some languages where, on the surface, this Law appears not to work. For example, in the Native American language Passamaquoddy, you can say things like the following:

nekom	nkisankumkunol	psite	Koluskap	utapakonol
he	sell	all	Koluskap's	cars

This Passamaquoddy sentence looks very similar to the English *He sold (me) all Koluskap's cars*. We've seen that, in English, *he* and *Koluskap* can't both be used to refer to the same person in sentences like this. The Pronoun-Name Law will block it. The Passamaquoddy example, however, can mean that Koluskap himself sold me all his cars, so in Passamaquoddy, the pronoun and name can both be used to refer to Koluskap. In fact, that's the most natural meaning of this sentence.[6]

Is Passamaquoddy, then, in violation of the Pronoun-Name Law? It turns out that it is not. The linguist Benjamin Bruening has looked closely at how this language works, and discovered that the Object in our example sentence is, in fact, not where it appears to be. Bruening analysed a wide range of complex sentence types in the language and came to the conclusion that the Object in this sentence, and sentences like it, is actually higher up than the Subject pronoun *he*. So even though the Object *psite Koluskap utapakonol* 'all Koluskap's cars' comes after the verb in terms of order, just as in English, it is much higher up in terms of structure.

The closest we can get to the Passamaquoddy structure in English is in a sentence like:

All Koluskap's cars, he sold me.

You can see that the phrase *all Koluskap's cars* is acting as the Object of the verb (in terms of meaning, the cars are being sold), but it's not in the normal position for an Object. Instead, it's at the front of the sentence. This isn't the most natural sentence of English, but, when you say it, you can use *he* and *Koluskap* to refer to the same person. That's because the structure looks like this:

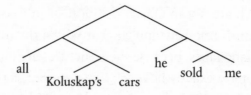

Tracing up from *Koluskap*, you can't reach *he* with one downwards step, which explains why the name and the pronoun can be used to refer to the same person.

The Passamaquoddy example isn't exactly like this English example (the order is different), but what's happening there is similar. The phrase *psite Koluskap utapakonol*, 'all Koluskap's cars' is acting as the Object of the verb, but is actually really higher up in the structure of the sentence, like this:

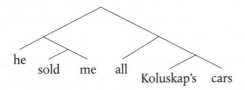

If you do the tracing your finger trick, you see that you can't connect *Koluskap* to the pronoun by going upwards from it followed by one downwards step (you have to go two steps down). Because of this, the structural connection between *he* and *Koluskap* can't be made. That means that the Pronoun-Name Law lets them refer to the same person in Passamaquoddy.

This example shows that we have to be careful when we apply the general Laws of Language to different languages. Just like physical or chemical laws, linguistic laws apply universally, but sometimes they are irrelevant for particular structures in particular languages. Laws of Physics that apply to gases are irrelevant for solids, because the structure is different. Laws of Language applying to names which are Objects, as in English, are irrelevant for names which are not in Object position, as in Passamaquoddy. The Pronoun-Name Law applies, but with a different outcome in the two languages, because the structures are different. We'll see in Chapter 9 exactly how an Object (or Subject) can appear in one position in one language and in another in a different language.

This means that Laws of Language are universal, in that they are part of Universal Grammar, but you can't necessarily see them at work in every language, because the structures of the language may not be relevant for particular Laws. No language will break a Law, but some sidestep them. Because the Laws of Language work as organizing principles, their effects are sometimes only detectable with careful analysis.

∞

I've argued that Analogy doesn't work as an explanation of how pronouns and names can be used refer to the same person. The Pronoun-Name Law gives us an alternative. Because human languages are built out of hierarchical structures, how we can build certain meanings is dependent on those structures.

Perhaps, though, the other part of Bybee's theory still holds. This is the idea that children learn that their languages are hierarchical through Chunking. Children do hear the same words together over and over again. Perhaps they begin to associate these words with each other, eventually coming to the idea that they form a chunk, and the undoubted hierarchy in human language emerges from that. This seems like a sensible approach. As children, we listen to what other people say, we notice patterns in their speech, and we use these patterns to organize how we speak our language. The approach predicts that children should start off not so great at language, and get better. At face value, this looks right. It also predicts that children shouldn't go beyond what they hear. They shouldn't come up with new patterns that they haven't heard before. Or if they do, those patterns should come from Analogy.

As you might have guessed, I disagree with this idea. All speakers of English, and many other languages, produce and understand sentences like the following:

Arnie's cat is Obama.

Arnie's sister's cat is Merkel.

The Subject of the first sentence is *Arnie's cat*, and that Subject has two bits to it. One is the noun *cat*, which is preceded by the name *Arnie*. In English, to express that Arnie owns the cat, we add the affix *'s* to it.

Let's think about this in terms of Chunking. As we hear people and the things they possess being referred to using a name, the *'s* affix and a noun, we associate these words, and abstract a general pattern from them. The chunked outcome would look something like this:

SOMEONE's SOMETHING
means *the something that someone possesses*

We might also hear examples like the second one above (*Arnie's sister's cat is Merkel*). In that sentence, it's not Arnie who owns the relevant cat, but Arnie's sister. If children hear sentences like these, they may well come up with another chunked outcome that looks like this:

SOMEONE's RELATION's SOMETHING
means *the something that a relation of someone possesses*

These two ways of Chunking up the data will capture these two sentences.

The linguist Avery Andrews has looked through a large collection of recordings of about 13.5 million words of adults talking to children.[7] Andrews found that, unsurprisingly, adults do say sentences like these. There were lots of examples of the *Arnie's cat* sort, and while there were far fewer examples of the second kind (*Arnie's sister's cat*), Andrews did find 82.

Andrews' 82 examples all fitted the second template quite exactly. Sometimes the middle noun was not a relation, but rather a pet, or a toy. Quite strikingly, in 74 of the examples the final noun was the word *name* (as in *Daddy's kitty's name*).

But now the problem arises. The chunks we've come up with are not adequate to explain the grammar of English. English allows you to multiply the number of people, or things, possessing each other without any obvious limit. We can easily expand our examples:

Arnie's sister's cat's name is Merkel.

Anson's friend's sister's cat's name is Merkel.

And we could go crazy:

Arnie's sister's neighbour's friend's cat's tail's tip's colour's brightness blinded me.

It is close to impossible to keep track of the possessive relationships in this last sentence. But we can tell there's nothing wrong with the grammar of it. Our syntactic ability gives us the structure, no problem. It's just that our memories aren't good enough to keep up with it.

Perhaps we can address this problem by adding in a third chunk to the ones we already proposed. Remember that there's no Chunking without repetition, so to do this, we'd want to be sure that a child had experiences of three nouns with an *'s* attached.

The problem for Chunking is that Andrews showed that children never hear three nouns of this sort in this kind of structure. In the 13.5 million words that Andrews looked through, there are no examples with the kind of structure that *Arnie's sister's cat's name* has. The maximum number of nouns followed by *'s* that the children would experience is two. Even those were very restricted and fairly rare.

The perfectly reasonable Chunking analysis we proposed above captures two nouns, but can't extend to more than two. This means that it isn't a good explanation, or indeed any kind of an explanation, of the grammatical capacities of speakers of English. The real syntax of English doesn't place any numerical limit on the number of possessors, though human memory capacity might do. If Chunking is right, this is a mystery. Chunking says you create chunks out of what you have heard, but children don't hear the kinds of structures that are clearly going to end up in their mental grammars. Why don't children grow up to be adults whose grammar of English is restricted in the way that Chunking would predict?

When we look across languages, we find something very similar. Languages, like German and the Amazonian language Pirahã, allow only one noun phrase in this kind of structure. Languages like English, or Hebrew, or Mandarin, allow as many as you like. What we don't find is languages that allow only two. This is completely unexpected if Chunking is right. Across the languages of the world, we'd expect to find at least some where exactly two possessing nouns is allowed, but no more. From the hundreds of languages that linguists have studied which use this kind of syntax to express possession, they either allow exactly one possessing noun, or they place no limit on the number of such nouns.

The linguistic experiences of English-speaking children don't involve more than two noun phrases with 's attached. Yet English grammar doesn't limit the number to two. Across all the languages of the world, we know of not a single one that has a limitation in this kind of structure to two. Chunking, though it sounds very reasonable, won't work. Chunking is not designed to allow an unlimited number of nouns or noun phrases in this kind of possessive structure. But an unlimited number is what the grammar of languages allows.

∞

Human language is rich, and flexible, and almost infinitely creative. That doesn't mean that anything goes. There are specific Laws that shape the structure and meanings of sentences. These Laws can't be reduced to general skills that allow us to match and extend patterns. The only way to explain the intricate organization of the grammar of a language is by Laws of Language interacting with the linguistic experience of speakers.

There are other arguments against Chunking and Analogy. If that is what is going on in language learning, why has English, over its history, not developed an order–based system to link names and pronouns? No special communicative efficiency seems to be got from structure as opposed to order. Surely order would be easier to learn. You'd just have a rule where you say the name first, then the pronoun, with no sensitivity to the structures of the sentences. But that is not, as we've just seen, how English works. It isn't the way that any language works, as far as we know. Languages almost always use structure as one aspect of what limits the links between pronouns and names. More generally, time and time again, in languages of the world, the order of words seems to play a minor role compared to structures they appear in.

Chunking also predicts that, at the earliest stages, children should be learning that specific words go with other specific words. So they might have a chunk built out of *a* and *cat*, but lack one built out of *a* and *dog*, just depending on what examples of language they have experienced. The psychologist Elena Lieven, and her co-workers Julian Pine and Ben Ambridge, have argued that that is what we see if we look at child speech.[8] A certain child might always say *a cat* and *the dog*, but never say *the cat* and *a dog*. This looks, at face value, like it supports Chunking, but, as so often in language research, we have to look deeper.

Virginia Valian, a language acquisition scientist at City University New York, has shown that children behave no differently from their mothers or other caregivers in this respect.[9] If we look carefully at what adults say to children, they also tend to use particular words together, but no one thinks that adults lack a grammatical rule that allows them to say both *a cat* and *the cat*. What people do say is different from what they can say. Valian has argued that children understand words like *a* and *the* when they are very young: they are able to distinguish them from nonsense words, and they know that they have a particular grammar. They don't learn them through Chunking, they create a grammatical rule from the start.

Connections in structure matter for meaning. Our rule about when pronouns and names can be used to refer to the same person is one example of this general principle. That rule is quite ubiquitous across languages. Children learning English seem to be sensitive to it, in just the same way that adults are, at a very young age. Experimental research since the 1980s has shown that quite young children (from three onwards) use the rule to interpret sentences. More recently, Jeff Lidz and his co-workers Megan Sutton and Michael Fetters at the University of Maryland have shown that children as young as 30 months old already use the Pronoun-Name Law.[10] Even more interestingly, they were able to measure the speed at which these children processed the sentences. They showed that the infants' ability to use the rule correlates, quite strictly, with how good they are at processing the abstract structure. So even very young children use abstract structure and Laws of Language.

The constructionists' proposal that linguistic structure and linguistic rules come from Analogy doesn't provide an adequate theory of language. Analogy should apply generally, but I've shown that it doesn't apply where it should. Analogy may be at play when language is acquired by children, but its effects are constrained by the way that the Laws of Language operate, by the kinds of abstract structure that we humans subconsciously impose on the language we hear around us.

The idea of Chunking also doesn't provide an adequate theory of language. Humans certainly chunk words into hierarchical structures, but we go beyond what Chunking allows. Chunking predicts human language should be limited in depth of the syntactic hierarchy: it should be as deep as it needs it to be to capture what we hear. Human syntax, however, doesn't place any restrictions on how deep hierarchy goes.

Kanzi is an extremely intelligent ape. He wants to communicate with his trainers and others, and has been surrounded by intensive language instruction, and linguistic interaction, every day of his life for decades. He does use his general intelligence to work out what people are saying, and to use symbols to express himself. Alia, when she was tested by Savage-Rumbaugh, was a little child. But she didn't use her general intelligence to complete the linguistic tasks she was set. She used her sense for linguistic structure.

There are other intelligences that can use language, artificial intelligences. These don't use systems of Chunking and Analogy, they use rich statistical abilities and huge data sets. Could human language work in the same way these do?

8

BOTLANG

In 2017, Facebook researchers had to react quickly to lurid headlines that spread around the world. The *New York Post* reported that 'Creepy Facebook bots talked to each other in secret language'. The UK's *The Sun* headlined with 'Robot intelligence is dangerous: Expert's warning after Facebook AIs develop their own language.' Within the article, they quoted a professor saying 'We do not know what these bots are saying. Once you have a bot that has the ability to do something physically, particularly military bots, this could be lethal.' Even the *International Business Times* asked 'Did we humans just create Frankenstein?'

What had happened? Were we really seeing the emergence of a secret language in which the robots would plot our downfall? Was the Terminator becoming a reality, and did it speak Botlang?

Not at all. But we were seeing something interesting about what a non-human language developed by computers might look like and how different that would look from human language.

What was behind all the hype? The researchers were, of course, not trying to create an artificial intelligence to supplant humans. They were interested in seeing whether something as everyday as negotiation could be automated. Negotiation is a pretty important skill. It's also a very human skill. It can involve making offers, but not really meaning them, answering questions, but not revealing the whole truth, asking for information, but really wanting to know something else. It's all about bluffing, blustering, compromising. It would be a tech giant's dream to automate it. But the human element makes this very hard to do. How is a computer program to know when best to withhold information, or when to leak lies. A computer might have a pretty good poker face, but would it be any good at poker?

The Facebook team tackled this question by first collecting over 5,000 examples of people chatting with each other online in a kind of negotiating game.[1] Each person was set the task of getting some objects (some hats, some balls, and some books). To do this, they had to make a deal with a counterpart, who wanted different objects. One person might type 'Can I have all three books?' Their respondent might reply 'You can have two of them, but only if you give me three balls'. Sometimes a deal would be made, and sometimes not.

The researchers took the transcripts of these negotiations between real people, and developed a computer program to learn from them. Their eventual goal was to create realistic interactions. Human users wouldn't realize that they were talking to a bot, and not a human. To work this through, part of their approach involved the bots interacting with each other.

The transcripts that the researchers collected of people negotiating were fed into the bots. The way that the bots learned from these was not how humans learn from the conversations they hear. The researchers programmed the bots to make connections between what kind of a deal was made and the conversations used to make the deal. They didn't programme in anything about the meanings of those conversations. The bots literally just saw the symbols (the letters and spaces that people typed) and connected these to good or bad deals. Unlike human negotiators, the bots knew nothing about what they were trying to do and nothing about the meanings of the words. They didn't understand the goal of the negotiation. They were just imitating, in a very sophisticated way, how humans acted and interacted.

The researchers 'rewarded' the bots when they made choices that led to successful outcomes and 'punished' them when they didn't. How do you reward a bot? You program another goal into the bot, a bit like a loyalty or reward card for a coffeeshop, and give it extra points when it does what you want. You also take away points when it doesn't. The bot is programmed to get as many points as possible. It's a bit like training a puppy with treats and stern words.

This technique was already familiar in Artificial Intelligence research: Google had used it to develop a famous computer system that defeated a grandmaster of the ancient, and infamously complex, Chinese strategy game Go. The Facebook team used this technique to improve the negotiating abilities of their bots. And it worked!

The bots succeeded in their negotiation tasks. They negotiated with each other, and agreed on the results of the negotiations. There was an unexpected outcome, though. Because the programmers had incentivized successful deal making, as opposed to keeping to the grammatical patterns of English, the bots, nick-named Alice and Bob, developed strategies for communicating with each other that made their conversations quite odd. Here's an example of their dialogue:

Bob: i can i i everything else
Alice: balls have zero to me to me to me to me to me to me to me to me to me
Bob: you i everything else

Alice: balls have a ball to me to me to me to me to me to me
to me
Bob: i i can i i i everything else
Alice: balls have a ball to me to me to me to me to me to

This is a bit disconcerting, it has to be said. The conversation doesn't look random. As Alice's number of *to mes* decreases (from nine to five), Bob's number of *is* increases. One of the Facebook researchers, Druv Bhatra, pointed out that this wasn't surprising. The bots were trying to achieve a goal, and they were coming up with ways to communicate about the situation that worked well for them. Apparently, repetition is a good way of signalling information in Botlang, but we don't find human languages working in this way. The tabloid headlines reported that the 'boffins' had to turn off the machine, as they were terrified that the bots were communicating in a way that no human could understand. In reality, this phase of the research was used to figure out how well the reward technique worked for negotiation. The researchers always meant to then reprogram the bots so that they'd keep closer to the grammatical sentences of English.

The bots did develop their own linguistic system to accomplish their tasks, a system that looks very different from human languages. Why is it so different? Could a human ever speak a bot language? And can bots do what humans can do with language?

When you ask *Alexa* to order you something from the internet, tell *Google Assistant* to show you somewhere on a map, or order *Siri* to open up your mail, it certainly looks like these programs are engaging with you. Perhaps not perfectly, perhaps a bit strained. For simple tasks, though, it seems that your computer is understanding what you are saying, and acting on it. Even a few years ago, this would have seemed the stuff of science fiction.

When I was an undergraduate in Edinburgh in the 1980s, I studied both Linguistics and Artificial Intelligence (AI). We could, at the time, build systems that would understand English, and interact with the world. In fact, one of the pioneers of AI, Terry Winograd, had built a programme in the late 1960s that you could interact with in a pretty sophisticated way.[2] Winograd's program, SHRDLU, read in instructions from a human in English, and performed simple actions. Its scope was very narrow. Winograd had designed a 'world' consisting of blocks with various colours and shapes. The program represented that world, and could change aspects of it. You typed in instructions, or asked questions, and the computer understood what you wanted, like this:

Person: Will you please stack up both of the red blocks and either a green cube or a pyramid?
Computer: OK.
Person: Which cube is sitting on the table?
Computer: THE LARGE GREEN ONE WHICH SUPPORTS THE RED PYRAMID.
Person: Is there a large block behind a pyramid?
Computer: YES, THREE OF THEM: A LARGE RED ONE, A LARGE GREEN CUBE, AND THE BLUE ONE.
Person: Put a small one onto the green cube which supports a pyramid.
Computer: OK.

This is a bit reminiscent of modern-day AI assistants like *Alexa* or *Siri*, but SHRDLU worked very differently.

Inside SHRDLU was a system that turned English sentences into the kinds of structure we have seen so far in this book.

SHRDLU drew on ideas from linguistics. It assigned abstract structures to sentences, then gave these structures meanings. This meant that it could understand and change the relationships between the blocks in its internal world. It could deal with the kinds of ambiguities we saw in the pot-dealer example in Chapter 2. It could understand how the parts of words come together to change the meaning of a whole word—like adding -s for Plural in English. It was sensitive to how sentences are built up differently, depending on whether they are questions or statements. Unlike the Facebook bots, SHRDLU linked the structures of the language it encountered with meaning.

SHRDLU was amazingly successful in its narrow domain, but it proved to be a huge undertaking to scale up beyond the blocks. SHRDLU relied on its programmer to code in all the information, and all the links between language and the world. Extending it beyond a toy world of blocks proved to be hugely expensive in terms of time and effort. On top of that, programs like SHRDLU proved very fragile. Today, you can typ[e in mispeled wrds on Gogle serch, and it'll know exactly what you want, just like a human does. SHRDLU just broke down!

More than that, it wasn't clear whether it was even possible in principle to get such a program to understand English as effectively as you or I do. Understanding the meanings of sentences is hard. Think about an example where we have a pronoun that is used to refer to the same person as a name. We've already explored examples where the syntax constrains how names and pronouns can be used to refer. But humans use much more information than that to figure out what's going on. Winograd used sentences like the following to make this clear:

The town councillors refused to give the demonstrators a permit because they feared violence.

Who does the pronoun *they* refer to here? The common-sense answer is that it refers to the councillors. But we can make one little change to the verb that *they* is the Subject of, and suddenly the pronoun most likely refers to the demonstrators:

The town councillors refused to give the demonstrators a permit because they advocated violence.

The problem here is that to work out who the pronoun *they* is referring to, you need to know a lot about politics, the likely actions of town councillors and demonstrators, and the meanings of verbs like *fear* and *advocate*. How on earth do you program all that knowledge into a machine? This kind of sentence highlights the difficulty of the problem. In fact, still today even the best AI language systems don't have anything like the performance of humans on such sentences.[3]

During the 1970s through to the early 1990s, huge amounts of effort were put into figuring how to develop a good enough understanding of human language to feed into AI-type programs like SHRDLU. An enormous amount was learned, both about human language and about how to develop computer techniques to analyse language. Winograd hand-crafted the language understanding system for SHRDLU. He built into SHRDLU a grammatical understanding of English. But, as we've seen, human languages are complex beasts, and we still don't understand everything about how they work. Winograd could only code in a fraction of the kinds of sentences people hear and say every day.

One big advance came when the process of coding in grammatical knowledge became automated during the 1990s. Using the huge amount of text that was becoming available on the internet and in collections of spoken and written records, hundreds of graduate students in linguistics were taught how to assign structures to tens of thousands of sentences. They created gigantic collections of sentences decorated with the treelike structures linguists use.[4] These huge repositories are called Treebanks, and are still in use today. Rather than directly programming computers with grammatical understanding, researchers developed computer programs that would scan the structures in the Treebanks, extracting grammatical information from these.

Many successful programs emerged from this. These contained rules distilled from the Treebanks of naturally occurring language. The idea was that this grammatical knowledge could then be used to understand examples of language that hadn't been encountered by the computer before. Linguists learned a lot about the complexities of English grammar from this, but developing working AIs from it proved extremely difficult. Programs, like SHRDLU, unlike humans, were fragile. They couldn't deal with ungrammatical, or mistyped, sentences. They baulked at words or structures they hadn't been programmed with. They broke, rather than bent.

The project of developing a SHRDLU-style computer system into something that could interact with humans, as though it was one of us, stalled. Partly, this was because of the difficulties of scale, expense, and fragility. Mainly, it was because something that worked better came along.

SHRDLU was a great example of what is called soft AI: we try to use what we understand about human intelligence to program an artificial intelligence. But there is another approach to AI: don't try to imitate humans, just use whatever you can to do the task. To understand language, don't concern yourself with grammar and meaning, or how humans process speech. Just use powerful computer systems, and vast amounts of data to get the job done. This is sometimes called hard AI.

All of the most successful Artificial Intelligence systems currently working, like *Siri, Google Assistant,* or *Alexa,* use the hard approach. Soft AI took a nosedive in the late 1980s and hard AI systems are now the norm. Many of these use a sophisticated version of the semantic soup idea we encountered when we talked about Kanzi. Kanzi used his intelligence to work out how the words he was exposed to were likely to be connected. He couldn't detect grammatical signals, but he was clever enough to work out what the words were likely to mean. The new AIs don't have Kanzi's intelligence, but they do have access to vast amounts of data. AI assistants, for example, pick up on the patterns of words in your question, then try to link them to patterns of words elsewhere on the internet. This is why you need to be connected to the internet to even use one of these AIs. Like Kanzi, they don't use the grammatical relations between the words. Unlike Kanzi, they look up huge data sets drawn from the whole of the internet. This allows them to figure out what your intentions were likely to be, without having to worry about the details of the grammar, or what you really intended. It's quite amazing, if you think about it: these hard AI systems appear to be rather good at knowing what we humans want, but they're doing this in a way that's profoundly different from the way that humans work.

For example, imagine you ask *Siri* 'Siri, set a reminder for me tomorrow at 9 a.m. for my meeting with Jill'. First some powerful speech analysers get to work to extract the words. They analyse the sound waves that you've just produced, and segment them into distinct words. This is an amazing accomplishment already. Then *Siri* sends the sentence it has worked out to Apple, who apply their language understanding systems to it. But these systems don't work like SHRDLU. You can see this by asking *Siri* the following 'Hey Siri, reminder Jill tomorrow

tomorrow tomorrow meeting'. *Siri* works out that you probably want to set a reminder for a meeting with Jill tomorrow, and it doesn't care about the grammar. You don't normally use ungrammatical sentences with AI assistants, which means that they seem human-like. But they're not. Not only do they involve sending information across oceans to servers possibly thousands of miles away, the way they solve the language understanding task is not how humans do it. They are an example of hard AI.

These systems may be hard AI, but they are also quite weak: they are specialized to particular tasks, rather than being generally intelligent. You can ask *Google Assistant* for something sensible, like 'How do I drive to Greenwich Pier from the Tate Modern?', and it will pop up a Maps application and show you. But if you ask how to swim from the Tate to Greenwich, *Google Assistant* doesn't really know how to respond, even though the Thames runs from one to the other, and you can get a riverboat! Humans would probably find the latter question easier to answer than the former (*Just swim eastwards down the Thames*). The AI's ability to think outside of the tasks it's programmed for is weak. A strong AI, on the other hand, would be general. A strong AI could do intelligent things in any situation. Our society today has lots of weak, hard AIs, but what would be really interesting is a strong, soft one!

These new types of AI do seem to have impressive powers to understand human language, but they're limited to working on particular problems, partly because they're not very good at grammar. This means they don't understand sentences like humans do. Could we make them better at being more like humans when it comes to processing language? Could we keep the advantages of these approaches, but have them be more sensitive to the rules of a language?

∞

One of the problems with the AI assistants is that they work mainly by a kind of key-word system. They hear or read particular words, and they look up those words, associating them with a likely meaning that they have, or action that's required. But human language doesn't work like that. Take a space with a bunch of words drawn on it like this, and make a sentence of English using all the words:

a

caught   cat

hungry   flying

a   mosquito

I'm guessing you went for *A hungry cat caught a flying mosquito*. But you're a creative human being, not a computer programmed to go for the most likely or useful interpretation. This means you could easily have gone for *A hungry mosquito caught a flying cat*. Or, *A hungry flying cat caught a mosquito*. There probably aren't too many flying cats around, and mosquitos usually don't catch other animals, but how the world actually is, or what people are likely to say, or have said before, is irrelevant. Syntax, the way that the combinations of words make meanings, doesn't care about probability or factuality. It's a fundamentally creative capacity. You can make new worlds of the imagination by combining words in ways they've never been combined before.

We know that this is because there are invisible relationships between the words in a sentence. These relationships structure the sentence, and give it the meaning that it has. That meaning is an automatic consequence of the syntactic structure. Flying cats don't exist, but syntax brings them into being, at least in the mind's eye.

Could we make the AI bots more sophisticated by paying attention to how words in a sentence are related, as opposed to just what words are present? This is what SHRDLU did. Modern AIs are successful because they use huge data sets, and probabilities to do their work, not rules of grammar. Could we use these abilities to learn grammatical rules?

We already met Claude Shannon back in Chapter 2 when we looked at his approach to communication. Shannon also came up with a way to think about the grammatical aspects of language, as part of his general theory. Shannon's proposal is compatible with how modern AI systems work. His idea was that each word in a sentence is connected to the words immediately before it by how likely they are to occur one after the other. Grammar is a side effect of these connections.

If you hear *cats like to*..., it's pretty unlikely that the next word is *happy*, but pretty likely that it's *play*. *Read* probably isn't going to be the next word, but you wouldn't be surprised if it were *sleep*. As we transition from one word to the next, the word we've just heard, or said, gives us very good clues to what's coming next. That transition from a past word to a future one is what's crucial.

Shannon's idea is that these patterns of probabilities are what give rise to the grammar of a language. If you have a particular word, then there are different probabilities for the various words that can follow it. The probability that *play* will occur after *cats like to*... is different from the probability that it will occur after *trees tend to*.... Because the transition between *to* and *happy* in *cats like to happy* is so improbable, we get the sense that that sequence of words is not a sentence of English. Shannon's proposal reduces syntax to statistics.

The probability of the next word in a sequence can be figured out by looking at what has come before. Let's take all the occurrences of the sequence *cats like* in English language books written in the 1990s. It turns out that of all the possible sequences of two English words, that sequence occurs 0.0000025% of the time—you can check this by playing around with Google's N-Gram viewer,[5] which allows you to track the use of words in texts across time. What about *like to*? That's much more frequent, as you might guess. It occurs as 0.005% of all two word sequences. And what about *to read*? Also frequent, at 0.003%. But *to happy* is even less likely than *cats like*, and *want to happy* doesn't appear at all. By looking at the likelihood of these two word, or longer, sequences, can we figure out which sequences are good approximations to English, and hence get computers to mimic the grammar of the language?

We can program computers to analyse text of various sorts, and tell us what the frequencies are. Once we have those, we can create new texts that match these frequencies. If you just look at pairs of words, the texts that are created don't look much like English. Here's a sample from an early attempt to do this, where each pair of words has the same frequency found in a collection of novels:[6]

> *you come through my appetite is that game since he lives in school is jumping*

The various pairs of words in this sequence appear at the frequencies that they appear in the original novels (*come through, through my, since he*, etc.). It definitely isn't English. However, if you expand beyond pairs of words to triples, or quadruples of words, the generated texts begin to sound more and more English-like. Here's a case where I've given quintuplets of words whose frequencies match the same set of novels:

> *road in the country was insane especially in dreary rooms where they have some books to buy for studying Greek*

What this means is that we can create approximations to English by looking just at the statistics of how words have already been used.

These ideas of Shannon's have proven pretty successful in developing AI applications to do machine translation. Google Translate, for example, used a sophisticated version of this idea for quite a few years, hooking it up to vast bilingual texts. Say you wanted to translate English to Spanish. The system looked up what translations had been given to all of the two-word pairs in your English sentence, then figured out which ones fitted best with the Spanish sentences it already had in its gigantic collection of Spanish data. The bigger the data set, the better the translations got. You can also improve this by looking at longer sequences of words, expanding beyond pairs to triples and quadruples.

However, no matter how big you make the sequences of words, you'll never capture the grammar of English, or Spanish, or any other language. As you increase the number of words in your sequence, you end up blocking out more and more sequences of words that are grammatically perfect, but just don't appear in your original texts. You can use this trick to build better automatic translation systems, but it won't get you to the grammar of a language.

We've seen from the beginning of this book that you need to go beyond what you've heard or read when you speak or understand a language. Human languages are open systems. We continually create new sentences to meet our needs, as we saw in Chapter 1. But AI applications working off Shannon's ideas, though undoubtedly impressive, just can't do that. This is why, if you used Google Translate for your Spanish homework until quite recently, it wouldn't necessarily improve your grade!

To understand the kind of problem that comes up, let's go back again to how verbs agree with Subjects in English. We've seen already that the verb in an English sentence doesn't agree with the noun that appears immediately before it:

*Many of the foxes that lived in my greenhouse last Winter were
frolicking at the back of the garden.*

The word *were* in this sentence matches its grammatical Number
not with *Winter*, or with *greenhouse*, but with *foxes*. If you're just
looking at the immediately preceding word, you'd expect *Winter
was*, not *Winter were*. If you base your machine translation system
on Shannon's idea, you're going to get the wrong result.

More recently, Google Translate, and other automatic translation systems, have started to use a new approach to get round
this problem. The new systems still work by looking at sequences
of words, but they incorporate a kind of memory. These AIs
basically remember the words they've already encountered, and
combine that memory with their knowledge of the likelihood of
combinations of words. In our example, the system remembers
it's read the word *foxes*, and when it gets to *were frolicking*, it looks
up the statistical frequency of the words *foxes* and *frolicking* occurring together. It also looks up the frequency of the words *Winter*
and *frolicking* appearing together. The former is more frequent
than the latter, and *foxes were* is more frequent than *Winter were*
in English in general. Putting all that together, the system works
out the most likely translation. Suddenly, everyone's Spanish
homework improves! In fact, this new version of AI translation
is so good that University Language departments have more or
less stopped giving translation homework. They either do translation under exam conditions, or they focus on how to use AI
translation systems to do the initial work that humans can then
improve on.

One of the challenges of these new systems is that their
computational innards are so complex that no one really quite
understands how they work. Researchers build a very general
system, then the system itself runs through vast amounts of

data, reconfiguring itself as it does so, until it has reached a point where it is trained. At that point, it's quite impossible to predict what it will do, or understand why it has done what it has done, but the results can be spectacular. These systems can recognize faces, translate extremely well, analyse building schematics, and do a host of other work that needs knowledge of complex patterns in data. Though the systems are doing an excellent job, no human really understands the details of what the system does at any one point. Indeed, this has become a major problem in using these new systems for tasks like expert diagnosis of medical conditions. They might do a lot better than human doctors, but if they go wrong, no one will know why. Almost as problematic, even if they go right, no one will be able to explain to the patient why they should take the machine's recommendation.

Tal Linzen, a researcher at Johns Hopkins, and his co-workers, have been using these AI systems as though they were partici-pants in experiments.[7] Researchers can play around with these systems comparing their performance at various language tasks to how well humans do in the same tasks. Linzen has shown that these systems get pretty good at doing Subject verb agreement of the *foxes were frolicking* type. In certain cases, with a bit of help, they can approach the level of humans.

Humans don't always get Subject verb agreement exactly right. There's a well known phenomenon, which happens when people are speaking, or reading quickly. It's called an Attraction Error. In Attraction Errors the verb is pronounced as though it's agree-ing with the word immediately before it, rather than with the grammatical Subject. Sentences like the following are sometimes produced:

The keys to the cabinet is on the table.

These sentences are occasionally spoken by people who would judge them to be an error. They meant to say *are*, and usually do say *are*, but just occasionally they make a mistake. Linguists usually think about Attraction Errors as little glitches that happen when your mind is planning out a sentence. It's a kind of slip of the tongue: you're planning to say one thing, but you say another. Fascinatingly, the sophisticated AI programs Linzen was working with also show Attraction Errors.

The twist, though, is that the errors are radically different to those that humans make. These two sentences are very similar in meaning, but have a different grammar:

The cats that are on the table are cute.

The cats on the table are cute.

Humans tend to make fewer Attraction Errors in the first type of sentence than in the second. That is, more people, more often, will substitute *is* for *are* in the second sentence than in the first.

One reason that has been suggested for this is that the first type of sentence has a more complex grammar. The Subject of the sentence (*the cats that are on the table*) contains what linguists call a relative clause. A relative clause is a sentence that can be used to add extra information to a noun. The second example doesn't have a relative clause in it. It has a simpler syntax that just involves a small phrase (*on the table*), rather than a sentence. The idea is that the extra structure in a relative clause makes the word *table* more distant from the verb, and less likely to affect it. Linzen and his team have shown, however, that sophisticated AI language understanding systems make far more errors in the first type of sentence. Exactly the opposite of humans. Even more bizarrely, the AIs made fewer mistakes than humans in the second type of sentence. They are better than we are in some cases, and worse in others.

Linzen's team also showed that the more nouns that come between the Subject noun and the verb, the more errors the computer programs make. The *foxes* example has two (*greenhouse* and *Winter*), while the *keys* example just has one (*cabinet*). Linzen's AIs perform worse and worse as the number of nouns increases. Humans don't do this. They are insensitive to the number of nouns. These machines are able to detect properties of the sentence invisible to humans. They might be mimicking human grammar using huge data sets, calculating complex probabilities, and even remembering what they've encountered before, but they are not doing what we are. Just like in the case of high-powered chess and Go programs, the computers are using quite different kinds of processes from those that we humans use. Human chess and Go grandmasters use highly abstract rules to guide their play. Computers crunch vast amounts of data. Human language users use abstract syntax to organize sentences. AIs use sequences of words and probabilities garnered from huge data sets. Unlike in the case of chess and Go, we are, at least for the present, still way better than computer systems at language. But who's to say that will remain true?

AI language understanding systems are, in some ways, more powerful than humans. These systems could easily learn a language whose rules are based on sequences. In fact, as we saw from the bots that opened this chapter, that seems to be the kind of language that emerges when we give bots free rein. But no human language works like this. Humans have had free rein linguistically for millennia, and the languages that have emerged all work on hierarchical, not sequential, principles. This is why the kinds of AI system that Linzen and his colleagues have been exploring don't make good models for human language. They have amazing capacities, but they are fundamentally a different kind of system from human minds. Once again, it's not about more oomph, it's about different oomph. These AIs have a sense of structure, but it's totally different from our human sense. They see visible sequences, we sense invisible structures.

9

MERGE

Take a deep breath.

As you breathe in, your lungs fill with air. The air is carried through every part of your lungs by tubes. These tubes are organized in a particular way. They branch off, one into the left lung, one into the right. The tubes fill our lungs by branching, branching, and branching again, into tinier and tinier tubes. Each branching point is similar, but not identical to the previous branch. Your breath, your very life, depends on this structure. It is a structure organized by the principle of self-similarity.

Self-similarity is everywhere in nature. Look at a fern: each fern leaf is composed of smaller replicas of itself, which are composed of yet smaller replicas. Or think of vast river deltas, where huge rivers branch out into smaller and smaller streams and rivulets until they vanish into the earth or oceans. Each branching of a river is similar to a previous branching that created that river.[1]

The internet has, without anyone overseeing it, evolved into a self-similar pattern, with huge hubs connecting to smaller ones, these themselves connecting, in just the same way, to yet smaller hubs all the way down to phones and laptops.

Self-similarity is everywhere because it is efficient. If a tube, developing into a lung, or frond into a fern, does the same thing each time it grows, then the genes don't need to specify the details of the growth. The same thing happens at the larger scale, and at the smaller. The same can be said for the branching of rivers. The same physical process operates. It makes no difference whether the river is the Amazon or a tiny stream. The self-similar structure of the internet shows that information spreads out in a similar way.

This principle of self-similarity, as we will see, is at the heart of syntax too.

All through this book, I've been arguing that the sentences and phrases of human languages, all human languages, have an inaudible and invisible hierarchical structure. When we are children, we impose this kind of structure on the sequences of sounds that we hear. Our minds can't understand the continuous blurring of sound in these sequences as meaningful language. Instead, as we saw in Chapter 6, we subconsciously chop them up into discrete bits: sounds and words. Bonobos and AIs treat sequences of words, either said or written, as sequences, not as hierarchies. Humans just don't seem to be able to do that. Syntax, the invisible hierarchical structure of phrases and sentences, is something our minds can't escape from. We hear or see language, but our minds think syntax.

The most basic units of language, words and word-parts, are limited. We can create new ones on the fly, if we need to, but we don't have a distinct word for every aspect of our existence. The number of words speakers know is a finite store. We can add words to that store, and we can forget words. But the sentences we can create, or understand, are unlimited in number. There is no store of them. Though we are only exposed to a finite number of sentences as we grow up, our syntactic abilities are not limited by these. All humans (without language disorders) develop, without any special training, to speak languages that make use of this syntactic ability. No animals, or AIs, do the same. These languages all combine words, or their meaningful parts, hierarchically. Different languages have different grammar, so the human syntactic ability must be both common to our species, and allow variation across languages. This ability must place no bound on the number of possible structures. It must allow the creation of a limited infinity.

The best candidate for such a process is what linguists call Merge, which is an example of a self-similar process. Merge is incredibly simple, but it is also quite specific. It takes two bits of language, say two words, and creates out of them another bit of language. Noam Chomsky has argued that Merge underlies the syntax of all human languages.[2]

For example, let's take two words, *drink* and *wine*. Merge says that we can take these two bits of language and from these create a new bit of language. But we don't do this by putting the two words in a sequence, like an Artificial Intelligence or Kanzi would do. Instead we create a new hierarchical unit. This unit puts together the verb *drink* with the noun *wine* to create the phrase *drink wine*. As we saw in Chapter 5, *wine* functions as the grammatical Object of *drink*.

Let's visualize how Merge works on a verb and its Object by putting things in boxes. Each box is a bit of language. If you're not in a box, you're not a bit of language. The things inside boxes aren't in a sequence; they don't have an order. The only information the box adds is that the bits inside it are grouped together. The words *drink* and *wine* are bits of language, and Merge says that a grouping of these can also be a bit of language. We can represent this with boxes like this:

Merge has created a self-similar structure: a larger bit of language containing two smaller bits of language. Boxes within boxes.

Now, in spoken or written language, we have to say one word then the other. That's just the way these channels of language work. We can't say the two words at the same time. Speaking, and to a lesser extent signing, flattens the hierarchy that Merge builds into a sequence, and that sequence has an order. This means that this structure, which is just one structure as far as Merge is concerned, can be pronounced in two ways. We either pronounce it as *wine drink*, or as *drink wine*. The grammatical Object either appears before the verb, or after it. Those are the only two logical possibilities. The first is the order we'd find in a language like Japanese, where we'd say *waino nomu*, literally *wine drink*. The second is, of course, English.

Linguists, as we've seen many times by now, usually write the outcome of Merge using little tree diagrams. These tree diagrams give us the information that comes from Merge (what words group together), plus information about order. The diagrams for English and Japanese look like this:

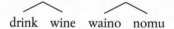

drink   wine   waino   nomu

These trees are the same in terms of Merge, but distinct in the order of their parts. We've translated the pure hierarchy that Merge gives us into a sequential order.

Let's assume that I drink wine. To express this thought as a sentence, we want to add a grammatical Subject to the bit of language we've just created. Since Merge says we can take two bits of language and create a new one, we can apply Merge to the word *I* and the linguistic unit we've just created:

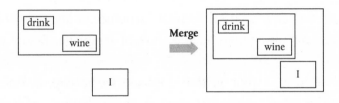

Now we have nested boxes. One big unit holding two smaller units, one of which is the result of a previous process of Merge. The word for *I*, in Japanese is *watashiwa*, so we can write our two trees like this:

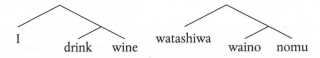

English is a Subject Verb Object language, as far as order goes, while Japanese is a Subject Object Verb language. But both languages Merge the verb and the Object first. In fact, the definition of grammatical Object is just the bit of language that is Merged with the verb.

There are also languages that mix up these orders. Malagasy, for example, has the English order for *drink wine*, but puts the Subject after that unit:

misotro   divay   aho
drink   wine   I

The Merge structure for Malagasy is just the same as that for English or Japanese, but while those languages put the Subject first, Malagasy puts it last. It is a Verb Object Subject language.

The last logical possibility in terms of what Merge will allow is where we say the equivalent of *wine drink I* to mean *I drink wine*.

This is pretty exotic in the world's languages. In fact, for many years, linguists were unable to find a language where that was the natural order to express this thought. However, a missionary-cum-linguist, the late Desmond Derbyshire, was, according to his friend Geoff Pullum, once lost in an Amazonian rain forest, and came across a tribe who had had little contact with the outside world, the Hixkaryana. The language family of the Hixkaryana, Carib, was known to linguists, but the Hixkaryana language was a new discovery. As Derbyshire worked on the language, he discovered that it had a basic order exactly the reverse of English.[3] To express *The boy caught a fish*, the Hixkaryana said the equivalent of *A fish caught the boy*:

kana   yanimno   biryekomo
fish   he-caught-it boy

Since Derbyshire's work, a number of other languages with this order, the Object Verb Subject order, have come to light, many in the same Carib language family, but some on other continents entirely. Even these rare languages can be thought of as having the same structure, given by Merge:

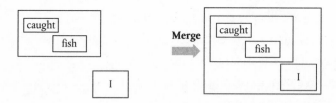

Hixkaryana has the Japanese order for the most deeply embedded unit, giving *kana yanimno* (literally *fish caught*). But it has the Malagasy order for the Subject *biryekomo*, 'boy'. The whole sentence can be given a tree diagram that looks like this:

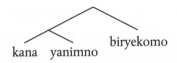

kana yanimno biryekomo

Merge isn't very complex, but it does a lot of what we need it to do. It applies to discrete units of language (words or their parts). It combines these, not sequentially, but hierarchically. It doesn't state what order the words have to be pronounced in, so it allows variation across languages. The hierarchical structure is the same in all four types of language we just looked at, but the order of the corresponding words is different. If we are talking Malagasy, where the Subject comes after the Object, or Japanese, where the Object comes before the verb, or even Hixkaryana, it will work just as well.

The fact that the hierarchical structure is the same allows us to express an important idea. The way that languages build up meaning is through Merge. Each Merge comes along with an effect on the meaning of the sentence, and that effect is generally both stable, and quite systematic, no matter what language you are looking at. That's why it makes sense to say that the Japanese and English sentences mean the same. They have different orders, but there is a deep commonality. Merge builds both structure and meaning in the same basic way in both kinds of language. Languages are deeply similar, not deeply different.

Merge is also open-ended. It both creates, and uses, bits of language. This means that Merge can reapply to something it has created. We can, in principle, create an infinite number of bits of language by using Merge.

Because Merge reuses its own output, it is a recursive process. Recursive processes are well known in mathematics, and form the foundation of modern theories of computing. Merge is a quite specific recursive process—it uses exactly two units, and it uses, and creates, structures which are linguistic. That's exactly what we want. If our human sense of linguistic structure is guided by Merge, that will explain why all languages are hierarchical and none are sequential. It also explains why human language is so unbounded, why sentences don't have a natural upper limit. Merge both creates, and limits, the infinite potential of language.

In any language there are some words that can be Merged together, and some that can't. In English, we can Merge *seem* and *happy* to create a unit, *seem happy*, and eventually a sentence like *You seem happy*. We can't, however, Merge *seem* and *sleeping*, to give *seem sleeping*. It's obvious what *You seem sleeping* means, and we could imagine it being said by a second language speaker of English. So Merge allows this, but English doesn't. In addition to Merge, we need to learn particular details of the grammar of our languages from what we hear around us growing up.

Merge looks a little like the constructionists' Chunking process. Both processes give you hierarchy. They are, however, deeply different. Merge imposes hierarchy on all it surveys. Chunking extracts hierarchy when the same things are experienced together over and over again.

We can see the difference by looking again at examples like *Arnie's sister's cat's tail*. We saw in Chapter 7 that English-speaking children don't hear examples like this—at least as far as we know from the millions of words of recordings we have from parents speaking to children. Chunking makes it mysterious why such phrases are perfectly grammatical in English. Why do children go beyond their linguistic experiences, developing a mental grammar that places no limits on possessive examples like this?

Let's look at how Merge would treat this. We say there's one bit of language *Arnie*, and another, *'s*. Merge puts them together to create a larger unit. The meaning associated with this is that the name is going to be related to or possess something else. Let's call structures with this form, and this meaning, Possessors, so we have an easy way of talking about the structure:

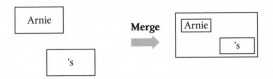

Now we use this Possessor we've just created, and Merge it with the noun *sister*:

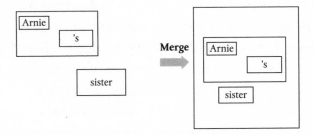

English word order puts Possessor first. Some languages, for example Scottish Gaelic, would put the Possessor last (*Arnie's sister* is, in Gaelic, *piùthair Arnie*, literally, 'sister Arnie'). The tree corresponding to the English order is:

This Merge process builds a meaning, just as we've done with every example of Merge so far. The meaning is built systematically from the meaning of the Possessor and the meaning of the noun *sister*.

Now we can see what the difference between Chunking and Merge is. If a child hears *Arnie's sister's cat*, and uses Chunking, that child can just come up with the Chunked outcome we saw in the last chapter.

SOMEONE's RELATION's SOMETHING
means *the something that a relation of someone possesses*

But Merge can't do this. It has to take two bits of language and put them together at every step. So what a child using Merge is forced to do is take the *Arnie's sister* unit, and Merge it with *'s*. Merge is just stupidly doing the same thing with *Arnie's sister* as it did with *Arnie*.

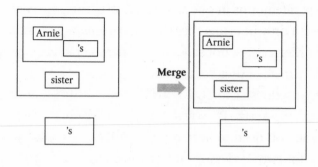

We already Merged a Possessor with a noun. We saw that with just the simple phrase *Arnie's sister*, when we Merged *Arnie's* and *sister*. We can therefore just do the same thing with the more complex Possessor we've just created. We Merge *Arnie's sister's* with *cat*. As a tree diagram, this looks as follows:

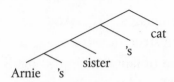

Just as *Arnie's* is a Possessor, *Arnie's sister's* is a Possessor. Possessors within Possessors. Merge automatically creates self-similar structures. There's no natural limit to this. Since we can have a Possessor as part of a Possessor, built by Merge, it follows that we can have that inside another Possessor and so on. The self-similar structures Merge builds have no end. They create unlimited language.

Merge is dumb. But it's very dumbness is what gives us an immediate explanation of why English, and most other languages, allow this kind of never-ending possessive. As a child you use Merge to impose structure on what you hear. If you hear *Arnie's sister's cat*, you are forced to analyse it as Merge creating a particular structure inside Merge using that same structure again. There's no other option. Merge won't allow you to put together three separate nouns in one sequential structure, because Merge is limited: it only puts together two bits of language at a time, and it creates hierarchies, not sequences. There's no choice but to get Possessors inside other Possessors.

Chunking doesn't force you to do this. With Chunking, you could just stop at two Possessors. You need specific evidence that shows you you need to keep going. With Merge, once you have evidence for two Possessors, infinity beckons.

Chunking would lead us to expect languages which allow exactly one Possessor or exactly two Possessors. This just doesn't happen. There are no languages which allow exactly two, but no more, Possessors in this kind of structure. The leap to the idea that there's no numerical limit on the number of Possessors doesn't come automatically with Chunking, but it does with Merge.

What of those languages, like German and Pirahã, which disallow this repeating structure. These languages allow only one Possessor. This kind of language is also predicted by Merge. In English, 's can be Merged with whole phrases. In German, the equivalent of English 's can only Merge to names and a few other things. This means that there's no evidence for a learner of German that a Possessor can be complex, so the German child doesn't have a syntax where Possessors can appear inside

Possessors. Merge, then, predicts that some languages can have exactly one Possessor, but once a child has evidence that two are possible, they know you can have an unbounded number.

Merge was proposed by Chomsky in the early 1990s, and was initially quite controversial. Chomsky's idea was that this single piece of mental technology, plus language specific constraints that children could learn from their linguistic experiences, was enough to capture the syntax of all human languages. The idea became important outside of linguistics when Chomsky coauthored a major article in *Science* in 2002 with Marc Hauser and Tecumseh Fitch.[4] That paper argued that having Merge is what really distinguishes human language from systems of animal communication. Merge, and how the structures Merge builds link to meaning and to the pronounced order of words, gives us the necessary core of human language. It underpins our creative use of language. It explains how, with just a finite number of words, we can express ourselves in such an unlimited way. Merge is Chomsky's proposal for how we can use language so creatively.

I said that the four types of languages we've just seen were the four logical possibilities given Merge. But there's an apparent problem. There are two other kinds of language: Verb Subject Object, and Object Subject Verb languages. In fact Verb Subject Object, is fairly common. It's the order found in Celtic languages like Scottish Gaelic, Mayan languages like Chol, and some Polynesian languages like Hawaiian. Let's look at Gaelic. In this language, to say 'A boy caught a fish', we say:

Ghlac	balach	iasg
caught	boy	fish

We can't get this order from Merge by just switching round the order of Object and Verb, or the order of the Subject with the rest of the sentence. As we've seen, there are only four possibilities when we do that:

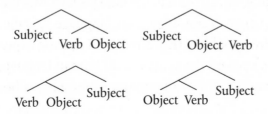

Merge doesn't look powerful enough to get us all the orders we want. Was Chomsky wrong to suggest Merge was sufficient as the single way that all syntax is constructed?

There are two ways out of this dilemma. The first approach says that, in some languages, we Merge the Subject and the verb together first, then add on the Object, like this:

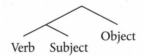

There is, however, a problem with this idea.

The notion of grammatical Object is closely connected to meaning. Not perfectly, but closely. Allowing this to vary across languages would mean that we lose any insight into why grammatical Objects in Gaelic mean the same kind of things as grammatical Objects do in other languages.

For example, Objects are never used to express the individual causing the action to take place. Take a verb meaning something like *bite*, where we clearly know who is doing the action, and who is suffering the consequences. In English, we could say *Lilly bit Pip*, which would look like this, with poor *Pip* as the Object:

Lilly   bit   Pip

But English doesn't have a verb which would work like the reverse of *bite*, where the grammatical Object would be the biter. We can invent such a verb. Let's call it *etib*. If you said *Lilly etibbed Pip*, it would mean that Pip did something—perhaps a playful nip—and Lilly suffered the consequences. Not only does English not have such a verb, we know of no language that has such a verb. The noun phrase Merged with the verb never has a meaning where the person it is used to refer to is doing the action.

This is easy to capture if we say that the meaning that Merge creates when a verb is Merged with a noun is consistent across languages. But if Gaelic has a different Merge structure, then we'd have to jettison that idea, and we have no explanation of the lack of verbs like *etib* across languages.

The second approach requires us to do nothing. Merge already has the power we need.

Let me explain how by looking more closely at Gaelic.

A first clue comes from the fact that verbs in Gaelic don't have a present Tense form. You can say *Ghlac balach iasg*, which is literally *Caught boy fish*, but you can't say the equivalent of *Catches boy fish*. Instead, Gaelic uses a separate word to express the present Tense, like this:

Tha	*balach*	*a'glacadh*	*iasg*
PRES	boy	catching	fish

That's the clue. The order after the word *tha* is Subject Verb Object. Just like English. We know that this order is easy to get via Merge. Gaelic really has two word orders—well, it has more, but two are important here: when the Tense is expressed by the verb (as in *ghlac*, which is past Tense), the order is Verb Subject Object; when the Tense is expressed separately, using the word *tha* for present Tense, the order is Tense Subject Verb Object. There's an alternation in the word order associated with Tense.

There's a very neat way of handling this using Merge. Remember Merge takes two bits of language, and creates a new bit of language. So far, we have just used Merge to take independent bits of language and put them together. That's how we got to the structure we've so far used for English, Japanese, Malagasy, and Hixkaryana.

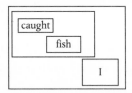

But Merge is recursive. It can go back and reuse something it's already used. This means we can Merge the bit of language containing the verb *caught* (that is, the box with *caught* inside it) with the bit of language containing everything else (that is, the large outside box). What we create is a new bit of language that contains the box with the verb in it, and the box containing that:

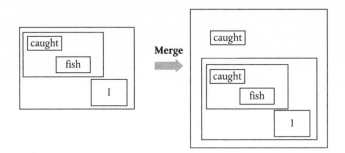

The verb *caught* here is both inside the box containing *caught fish* and inside a larger box that contains *caught* and a box containing everything else. It is in two places at once. One bit of language that appears in two places in syntactic structure. We've created a kind of loop in the structure.[5]

We didn't have to extend Merge to do this. Merge is incredibly simple—take two bits of language and create a new bit of language containing them. In fact it's hard to think of any simpler way of creating hierarchy out of objects. Merge just has the consequence that we can reuse bits of language that have already been used, in the way we just saw with *caught*.

The question now is whether this will give us a way of understanding Verb Subject Object languages like Gaelic. If I write this as a tree diagram, putting things in the right order, we can begin to see how this side effect of Merge gives us a solution to our dilemma:

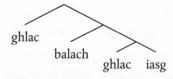

The Subject here, *balach*, is 'boy' and the Object *iasg* is 'fish'. The verb *ghlac*, 'caught', Merges with its Object and gives the meaning associated with catching fish. Then the result of that process Merges with the Subject. This can be ordered as Subject Verb Object. Finally, Merge applies to a bit of language it has already used (the verb *ghlac*) and Merges that with the Subject Verb Object unit. We can now pronounce this tree like this:

Ghlac	*balach*	*ghlac*	*iasg*
caught	boy	caught	fish

Oops! This isn't quite what we want, since the verb (*ghlac*), 'caught', appears twice. That verb is just one thing, though it appears in two places, because we've reused it.

The word *ghlac* appears in two places, but it is just one bit of language. So far, where we have one bit of language it is just pronounced once. This is a basic Law of Language:[6]

Law of Pronunciation
Each bit of language is pronounced just once.

Our previous Law of Language (the Pronoun-Name Law) expresses how a syntactic structure links to meaning. The Law of Pronunciation expresses how a syntactic structure is related to sound or sign. Both are part of Universal Grammar: Merge creates the structures that the Laws connect to sound and meaning. The mental grammar of a language like English, in a finite way, expresses this link between sound and meaning over a potentially infinite number of sentences.

With the Law of Pronunciation in place, we can pronounce our Merge structure and get the right sequence for Gaelic: *Ghlac balach iasg*, which is *Caught boy fish*. The word for fish, *iasg*, is still the Object of the verb, since it is in a box with the verb. But we've reused the verb, and it's in that reused position that we pronounce it.

We've not complicated how Merge works to do this. Merge just says to take two bits of language and create another. Merge can use things it's already used, i.e. it is recursive. It automatically gives us a way of treating Gaelic and other languages where the verb comes before the Subject.

This idea also makes sense of what happens in the present Tense in Gaelic. In the present Tense, we don't reuse the verb, we just Merge in the word *tha*, that conveys present Tense:

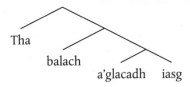

The rest of the structure simply follows the Subject Verb Object order, like English, or Swahili, or hundreds of other languages.

There is more to say, of course, about how to analyse Gaelic sentences. Why, for example, when we Merge in *tha*, does the verb appear as *a'glacadh*? Why does *tha* Merge only once the Subject has Merged? The analysis here only scratches the surface.[7] But the basics are right.[8]

When we use Merge to add in the present Tense word, or to place the verb in the outermost box, we are doing something very similar in both bases. Merging *tha* expresses present Tense. Merging the verb expresses past Tense. Both cases of Merge are adding in information about grammatical Tense.

Linguists have argued that many languages use this application of Merge, the one that adds on information to the basic Subject Verb Object unit, to introduce Tense information. We've seen that the first application of Merge is linked to the meaning that a verb sets up with its Object. The second application links to the meaning that the verb sets up with its Subject. The third Merge adds in information about Tense in Gaelic, as we've seen. We can see Tense information added in in English too. For example, in sentences where we emphasize what's going on, English uses a special verb *do*, and this carries Tense marking:

Lilly does catch fish.

The word *does* here expresses both emphasis and Tense. Is this similar to Gaelic? It looks a little different in word order. The Subject in the English sentence appears before Tense (*does*), but in Gaelic, the Subject appears after Tense (*tha*).

Merge provides us with an elegant analysis of the differences between English and Gaelic. The idea is that both English and Gaelic add in Tense information to the unit containing the Subject Verb and Object. The two languages work just the same as far as the relation between Tense and the rest of the structure goes. The meaning associated with Merge is the same in both languages. The difference is that Gaelic either Merges *tha*, or reuses the verb, while English reuses the Subject:

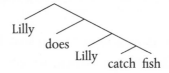

Since *Lilly* is just one bit of language, it is pronounced just once, where Merge has reused it. The tree diagram is therefore pronounced as:

Lilly does catch fish.

There are lots of questions that this opens up. Can you re-reuse a bit of language? Do languages differ in how they pronounce reused bits of languages? Are there limits on what can be reused? Are there further Merge operations that extend the meaning of the sentence? What does this approach predict of related data, and is it right? There are answers to these questions, but they take us into details of linguistics that go beyond this book.

I'm going to give one final example of where Merge reuses a bit of language. We met, back in Chapter 1, the idea that different languages pronounce different question words, like *what, who,* or *where* in different places. English, and many other languages, like to place these words at the start of the sentence:

What did you say that Lilly caught?

In this example, the word *what* is the Object of *caught*. Lilly caught something. But *what* doesn't appear next to the verb. It appears at the start of the sentence. In Mandarin Chinese, in contrast, the word for 'what' just appears where you'd expect it, right next to the verb it is the Object of:

Nǐ	shuō	Lìlì	zhuā	shénme
You	say	Lilly	catch	what

Here *shénme*, which means 'what', comes immediately after the verb *zhuā*, 'catch'. This is the normal position for Objects of verbs in Mandarin. In English, in contrast, the question word, although it is the Object of the verb in terms of its meaning, appears at the start of the sentence. The two languages differ in that English displaces the word *what* to the start of the sentence, while Mandarin leaves it where it is.

Merge gives us a simple way of understanding the differences between Mandarin and English. In both languages, the verb Merges with the word for 'what', so that word is the Object of the verb. In Mandarin, the Object is just left in peace. In English, however, it is reused. First it is Merged with the verb. Once lots more Merges have built up the rest of the sentence, *what* is reused. It is Merged with the whole sentence. Just like the other cases we have seen, *what* is pronounced where it has been reused. So although *what* appears in two places in the structure, it is just pronounced once.

Children learning these two languages use Merge to connect the verb and its Object in just the same way. But children learning English hear the question word not where they expect it, but at the start of the sentence. They therefore add information to their developing unconscious syntax of English: reuse a question word once the sentence has been built up. Again, lots of questions arise. What happens when there's more than one question word? Do all question words work the same? Do Object question words and Subject question words work in the same way? How much does the child have to learn, and how much does Merge itself, or other Laws of Language, guide the process? Again, I'll put these questions aside, as they take us into details of linguistic theory.

∞

Merge allows us to express the deep commonalities between languages. Not only are all languages hierarchical, there is a way of thinking of them as all fundamentally the same. Languages build the fundamental meaning relationships between bits of language in much the same way. Verbs Merge with nouns, or phrases containing nouns, giving a verb-Object relationship, with a consistent impact on meaning. That unit is then Merged with another noun or phrase, which gives the meanings associated with Subjects.

In the examples we have seen, Merge then adds in Tense information. Is this universal? The linguists Betsy Ritter and Martina Wiltschko have argued that there is a universal way of characterizing this application of Merge, but it doesn't always involve Tense.[9] Ritter and Wiltschko argue that this application of Merge always adds, to the meaning of the sentence, information about the situation that the speaker of the sentence is in.

Tense is one way of doing this. It links the situation described by the sentence to the time that the speaker says the sentence—has it already happened? Is it going to happen? Other languages anchor the sentence to the speaker's situation in different ways.

For example, Halkomelem Salish, a Native American language spoken on the west coast of Canada, doesn't express Tense in its sentences. Instead, it obligatorily expresses the location of the event in terms of distance from the speaker. The following two Salish sentences have exactly the same meaning in terms of when the event is happening. They are compatible with it having happened before the sentence is said, or with it happening while the sentence is said. But they differ in location, and that location has to be expressed for the sentence to be good Salish:

í	qw'eyílex	tí-tl'ò
here	dance	he

lí	qw'eyílex	tí-tl'ò
there	dance	he

The difference between í and lí is that the speaker is close to the dancing when they say í, and further away from it when they say lí.

The Salish examples suggest that, when Merge applies to the unit containing the Subject Verb and Object, it adds a meaning that links what the sentence is talking about to certain aspects of the act of speaking. Not any aspect. The time of the utterance matters, as we see in English. The location of the speaker matters, as we see in Salish. But not what the weather is, or whether the speaker is worried, or hungry, or whether there is immediate danger, or a thousand other relevant aspects of human lives. As we've seen, language is picky. The meanings associated with Merge come from a small set of possible meanings. The human mind limits what gets into language.

Merge is a single, simple, rule, that can create the endless structures that we find in human languages, linking these to meaning.

There is, as far as Merge is concerned, one human language. We all speak dialects of it. We use different words, we order those words in different ways, and some languages reuse bits of language where others don't. But the basic building blocks are the same, the meanings that are associated with Merge are taken from the same small pool, and Merge itself is universal to humanity.

10

GRAMMAR AND CULTURE

The way we speak, or sign, is a profound expression of who we are. The words we use, how we pronounce them, how we put them together, communicates far more about us than the literal meaning of the sentences we utter. Language is not just about structure, it's also about expressing our identities, and it binds together cultures and sub-cultures.

The approach to thinking about language I've sketched out over the last nine chapters seems very far removed from this. That approach says that at the core of human languages, all human languages, is a piece of mental technology, Merge. Merge works on discrete units of language, and builds new discrete units out of them. It creates hierarchy, not sequences. Merge works its ultimate magic through the self-similar structures it builds. It also opens up a way that languages can vary, as we saw when we looked at the differences between Gaelic and Hixkaryana, or English and Chinese.

Merge gives us the wherewithal to use language creatively. But it does so in a limited way: it sees the discrete, the hierarchical, and the universal grammatical links to meaning. But what about the continuous, the sequential, and the contextual or cultural links to meaning?

∞

We saw in Chapter 6 that humans, and other animals, impose discrete categorical boundaries on sounds, even though these sounds are physically continuous. Does that mean that we are literally incapable of hearing the continuous differences? No. It simply means that we don't co-opt these differences into the linguistic system of sounds that make up our languages. But that leaves a lot of information that can be used for other purposes.

One of the most famous examples of this comes from research by Penny Eckert, a linguist based in California who is interested in social aspects of language use. In the late 1980s, Eckert studied the way that language was used in a High School in the Detroit sub-urbs.[1] Vowels are one of the most continuous parts of language, and they change, sometimes quite radically, across dialects, and also across time. Eckert showed that this continuous property of words was used by different groups of schoolchildren to signify their identity as belonging to different High School tribes: Jocks or Burnouts.

Jocks are keen on sports and school activities. They feel part of the school, and respect authority. Burnouts are anti-school, and anti-authority. They drink, they smoke, and they do other things frowned on by the school. Eckert studied many aspects of the speech of these groups, but we will just consider one here, the vowel that the students used in words like *fight*, *light*, and *sight*.

That vowel is being affected by language change across a large swathe of North America. It's a change that is strongest in the cities, and weaker in the suburbs and countryside. The change makes the vowel in *fight* and similar words sound as though the word was really *foyt*. You can hear it for yourself if you say *fight* very slowly, and, as you say the vowel, you pull your tongue up and back slightly. As you move your tongue around in the mouth, the sound of the vowel continuously changes. Eckert found that Burnouts were unconsciously pulling their tongues further up and further back, and they were doing it more often, than the Jocks. The Burnouts associated that sound with the city, and with the people who lived there: street smart, autonomous, anti-authority. They were using the continuous nature of the vowel to link themselves with people in that social group. In fact, Eckert showed that the most burnt-out of the Burnouts, those who were wildest in their behaviour and most alienated from school, were also the most extreme in how frequently and strongly they pulled that vowel up and back when they spoke. The continuous nature of the vowel space links to the continuous nature of the school children's social identities. Eckert calls this the social meaning of the vowel. Social meaning is quite distinct from grammatical meaning. I can change the vowel in *catch*, when I make it past Tense, and it becomes *caught*, or the vowel in *man*, when I make it Plural, to *men*. These are discrete vowel changes that are connected to grammatical meanings (Tense and Number). Eckert pointed out that continuous changes can connect to continuously changing social meanings.

Social meaning also attaches to other continuously made sounds.

Say the sound *sssssss*, like a snake hissing. Feel where your tongue is in your mouth. Now pull it back slightly, and hear that the *sssssss* gets more turbulent, slightly lower in pitch. Pull it back further, and the *sssssss* becomes more like a hushing noise, *shshshshshshsh*. Now push your tongue forward again. The *sssssss* will, for some people become sharper, maybe even slightly whistley. That *s* sound is still *s*, but you can change how you make it, and how it sounds.

Normally, English and other languages treat *s* as a single discrete unit of sound, that fits into a system, like we saw in Warig in Chapter 5. A single sound can entirely lack meaning. In a word like *sue*, all the *s* does is arbitrarily contribute to the form of the word, distinguishing it in pronunciation from the word *zoo* through voicing, or *too* through its nature as a fricative (where the air gets through the mouth) rather than a stop (where it momentarily doesn't). In a word like *cats*, *s* adds grammatical meaning, the number of cats we're talking about is more than one. In both of these cases the contribution of *s* is discrete. In the first, it distinguishes words, in the second, it adds a unit of meaning. When you say *sue* or *cats*, you can say the *s* with a high hissing pitch, or with a lower looser sound, and it makes no difference to the function in the language. Those continuous aspects of how we produce *s* do, however, carry a different kind of meaning, social meaning.

One of the social meanings attached to *s* has been how it is perceived to connect to sexuality, especially for men. David Sedaris, the gay American comedian and essayist, tells in his story 'Go Carolina' how he was dragged to a speech therapist to sort out his pronunciation of that letter. 'Plurals,' he writes, 'presented a considerable problem ... Possessives were a similar headache.' Sedaris writes that he pronounced his own name 'thedarith', and the job of his speech therapist was to sort out the boys with that problem. 'None of the therapy students were girls,' he writes. 'They were all boys like me who kept movie star scrapbooks and made their own curtains.'

Sedaris's essay is a humorous piece about homophobia, both social and internalized. But does the social meaning that he wrote about have any basis in fact? Do gay men use the continuous aspects of the pronunciation of *s* as a way of unconsciously signalling that part of their identity?

Ben Munson is a phonetician, someone who works on the details of the way we produce and hear sounds. In 2006, he carried out a study at the University of Minnesota. Munson asked self-identified straight and gay participants from Minnesota to record a number of words.[2] When he analysed the detailed phonetics of how the various sounds were made, the differences between how straight and gay men pronounced *s* were extremely subtle, but they were there. Gay men tended to pronounce the *s* sound slightly more forward in the mouth than straight men did. The differences that Munson detected are close to impossible to hear when you consciously listen to the *s* sound. He used the kind of spectrogram technology we saw in Chapter 6, combined with sophisticated statistics to analyse them. Nevertheless, when Munson tested a different set of individuals, asking them to guess the self-identified sexuality of those who had made the recordings, people did a fairly good job overall on getting the sexuality right. They were subconsciously picking up on very subtle cues in the speech they were listening to.

It's important to say that Munson's findings were about the overall group. Particular individuals might identify as straight, but have more forward *s* sounds, and vice versa. So this isn't a result about individuals. It appears to be a very subtle shift, detectable by looking across the whole group, where men who identify as gay have a subtly different way of saying *s* from men who identify as straight.

Munson's findings were also interesting, as they showed that the particular pronunciation of the *s* sound by gay men tended to connect, in peoples' judgments, to the clarity of the speaker. The *s* sound that the gay men in the study made was, if anything, clearer and more distinct and more *s*-like than the *s* sounds made by straight men. Sedaris certainly didn't need a speech therapist!

This subtle, unconscious relationship between self-identified sexuality and the pronunciation of s is something which is not built into the sound s in any way. It's not connected to physical anatomy. It's a kind of meaning that signals belonging and connection to a community. Different communities have subconsciously used the extra information that the continuous differences in frequency of s provides to express particular meanings that are socially important to them. The more forward pronunciation of the s sound for some people signifies an identity, but it can also signal more nuanced meanings than that.

We can see this from a study done in Shasta County in California by Rob Podesva and Janneke Van Hofwegen in 2016.[3] These researchers compared gay mens' pronunciations of s in Shasta County, a conservative, fairly rural, part of northern California, with earlier work that had been done in San Francisco. They discovered that, though gay men in Shasta County tend to make the s sound more forward than the straight men there, their s sounds are pronounced further back than gay men in the city. Podesva and Van Hofwegen argue that these men are indeed taking advantage of the continuous nature of s to signify their identity as gay men, but they do so in a way that takes account of the more conservative area they live in. The conservative rural identity of the county is connected to an s sound that is made further back in the mouth. Gay men in Shasta County are carefully, but subconsciously, negotiating a complex set of social meanings in their pronunciation of this sound.

In Chapter 5, I argued that no language uses the length of a vowel to continuously mark how far back in the past the action described by a verb takes place. Grammatical meanings, like past, Plural, Mirativity, and so on, are linked to the discrete categories that Merge manipulates. Social meanings, like identity, affiliation, class, can be linked to continuous aspects of language.

There's a division of labour in language, but both linguistic meaning and social meaning are fundamental parts of the system that we use in our day to day lives.

Because Merge is blind to anything but the discrete, it leaves the way open for the rich information in speech and sign to be used to convey non-grammatical meaning. The example I just discussed concerned sexuality, but almost every aspect of our cultural and social lives can be connected to the aspects of language that Merge cannot access.

A fascinating example comes from the work of Jennifer Smith, and her team in Glasgow. Smith has, for many years, been working on the speech of Buckie, the rural Scottish town where she grew up. Part of her work studies how young children learn the dialect of their community, and how they distinguish it from the more widely spoken variety of English they hear on TV, or on the radio, or the internet. Smith's team has built up a large collection of recordings of children with their mothers or other caregivers. This data lets us see exactly how certain social contexts influence the kinds of grammar speakers use.

Here's an example of a mother, Lesley, chatting to her daughters, Lucy and Beatrice. The names are pseudonyms to maintain anonymity. I've altered some of the spelling, and standardized some of the Scots words, to make the conversation easier to understand:[4]

> Lesley: Are you going to get your frock on and we'll go up to
>   Bob's and get this mince?
> Lucy: No but we're watching the wee—the video.
> Lesley: Well can we not do it when we come back down?

Lucy: Okay then.
Lesley: Get your frock on then.
Lucy: *Dona* know where it's at.
Lesley: Where did you put it?
Lucy: I *na* know.
Beatrice: It's down there.
Lesley: I see it. You're sneaky. You was hiding it from me.
Lucy: I *dona* know where it's at now.
Lesley: There it goes.
Lucy: I *dona* know where it's at.

Later in the conversation we hear Lucy arguing with her sister Beatrice. Beatrice insists that she knows where Bob's place is. Lucy's response is:

Lucy: No, you *do not*.

Lucy has made her sentences negative in three different ways: she says *dona*, *do not*, and just *na*. These three versions of sentence negation are also what adult Buckie speakers use, showing that Lucy, who's just over three years old, has already mastered this aspect of the dialect of her community.

Now adult speakers from Buckie use these different forms more or less frequently. Smith also has a corpus of conversations among adult speakers. When she analysed how negation is expressed, she found that adults use *dona* and *na* about equal amounts of the time, but rarely use the standard version *do not* (or *don't*). But when the caregivers of the children, who are just the adults in the community, are speaking to the children, they use the standard version much more (about one third of the time). The children use it even more (two thirds of the time). The children's caregivers are changing the frequency of the different ways they make sentences negative to perform a specific social task: talking to their children rather than to their friends.

So far we might explain this by saying that mothers want to use the standard form more with their children, helping them to 'talk properly'. But, Smith showed, something more than this is going on.

Interactions with children can be playful and intimate, but they can also involve teaching, and, quite often discipline. Smith's team separated out the contexts in their recordings of when the children were being scolded from when they were playing. It became very clear that the adults were using the standard form *do not* or *don't* far more in the scolding situations, and the children were following their lead. In contexts of play or intimacy, the dialect versions *dona* and *na* were used more frequently. The adults were associating the standard form with more formal situations, and the dialect forms with intimacy and play, and the children were picking that up from an early age.

Merge, of course, doesn't pay attention to frequencies. But just as we saw with the continuous aspects of sound, the frequency or rarity of particular grammatical forms—in this case, ways of making a sentence negative—can be used to signal social meanings, to create a difference between formal and informal ways of speaking. Language again makes the same division: combining discrete, grammatical elements together gives linguistic meaning (is the sentence negative or not?), while meaning attached to the social context is signalled by how often particular ways of expressing that meaning are used in a conversation (is this conversation formal, or intimate?).

∞

There's an urban myth about Bob Geldof, the lead singer of the 1980s band The Boomtown Rats. It goes like this. During the Live Aid campaign to raise money for the 1985 famine in Ethiopia, Bob Geldof was on live television. It was a long time ago, so there was no internet, no online giving. People had to call in and pledge money and it appeared to Geldof that that wasn't happening quickly enough. Geldof was being interviewed by a journalist, who asked what people could do to help. A frustrated and impassioned Geldof grabbed the mike from the journalist, leaned into the camera, and said 'Just give us your fucking money!'

Let's think for a second about what Geldof is reported to have said—the actuality is disputed, but the example is relevant. The word *fucking* here doesn't contribute a sexual meaning. It expresses Geldof's passion, his frustration, his impatience. In fact, our imaginary version of Geldof could have said the following— apologies for the swearing, this section is going to have a lot of it:

Fucking just fucking give us the fucking money!

If he had done so, using a sequence of *fuckings*, what is expressed is just more of the same. It's more impassioned, more frustrated. When you add an extra *fucking* to the sequence, the meaning it contributes isn't like the meaning that, say, *winning* contributes in a sentence like the following:

Just give me the winning ticket!

The word *winning* can't be added in a number of times. This sentence is just weird:

Winning just winning give me the winning ticket!

The linguist Chris Potts has shown that certain words add in what he calls expressive meaning.[5] This is a different dimension of meaning from the purely linguistic, and from the social. We use swear words like *damn*, *bastard*, and *fucking* to express an attitude to the whole of what we say. We can interpolate these words in a sequence throughout the sentence. In fact, we can even interrupt single words with them. My favourite example is from the TV show *Absolutely Fabulous*, where the main character, Edina, is complaining that her daughter, Saffy, is going to leave home to go to university:

> Edina: *Where is it you're going?*
> Saffy: *Aberdeen.*
> Edina: *Aberdeen*
> Edina: *Aber-bloody-deen*
> Edina: *I don't know anybody in Aber-bloody-deen, darling!*

Potts argues that expressive meanings are fundamentally about the speaker's perspective on the current situation. They aren't about the truth of what is said. Remove Edina's *bloody*, and it makes no difference to what she's saying, though it does affect what she's communicating. Expressive meanings are disconnected from grammatical meanings like Tense. The following sentence is in the past Tense, but the expressive meaning contributed by *fucking* connects to how the speaker feels, right now, about the person they are referring to:

> *Our fucking prime minister made a huge mistake last week.*

Expressive meaning is valid for the speaker at the moment of speech. In a sense, it is like a gesture. When you frown as you speak, that frown conveys how you feel, right now. It's disconnected from other people, other times, and other places.

There's interesting evidence that expressive items escape certain syntactic rules in languages. The following grammatical construction in English seems to require that what comes before and after *or no* must be the same:

> *Water or no water, I'm not crossing the desert.*

If we try to make them different, the sentences are pretty strange:

Water or no H$_2$O, I'm not crossing the desert.

Water or no fresh water, I'm not crossing the desert.

When we use an expressive word, though, the sentence is fine:

Water or no fucking water, I'm not crossing the desert.

This suggests that the way that these expressions integrate into the sentence is different from other kinds of words. We saw that we can repeat *fucking*, interspersing it through the sentence, and even in the middle of words, and we don't get a sense of redundancy when we do that. The expressive meaning simply intensifies. But an expressive word can't go anywhere. I can say *my cute cat* and also *My cat is cute*, but though *my fucking car* expresses an attitude to my car—perhaps it's broken down for the umpteenth time—*My car is fucking* just doesn't have the same meaning. Words that add in expressive meanings seem to work in a slightly different dimension of grammar than other words do. They occur in sequences, can repeat in those sequences, and can interrupt the middle of other words. Expressive meaning takes advantage of the sequential nature of the speech stream, quite a different dimension from the hierarchies of phrases constructed by Merge.

There are other dimensions of meaning than the grammatical one. Merge combines discrete bits of language and builds grammatical meanings with them via hierarchical structures (who did what to whom? When did it happen?). But social meanings can be signalled by the continuous aspects of sound or the frequencies of particular grammatical choices, while, expressive meanings are expressed by interweaving certain words as a sequence throughout the pronunciation of a sentence.

∞

Merge builds certain kinds of meaning from a finite set of words and a limited set of grammatical properties—we've seen noun, verb, Tense, Number, Person, case, location, Evidentiality, Mirativity, but there are quite a few others. We've seen different languages use different properties. All languages we know of seem to have a noun verb distinction, but not all languages have Tense. Whether a language has a particular grammatical meaning is not determined by Merge or by any Laws of Language. How, then, do they get into a language in the first place? One way, it turns out, is through the social ways that people interact with each other.

Inner London is a massively multilingual city. Hundreds of languages are brought to the city by immigrants, and the children of these immigrants grow up often speaking one or two languages at home. This can mean that their exposure to English is limited. It comes mainly at school, through media like TV, YouTube, and radio, or through their friends. Many of their friends will be in a similar situation, though with different home languages (Yoruba, not Sylheti, Malay, not Turkish). This means how the children learn English is similar to what we saw in the Nicaraguan Sign Language case discussed in Chapter 4. People from incompatible linguistic backgrounds are brought together, and have to work out a new way of communicating.

This complex linguistic situation has led to the emergence of a new, changing, variety of English. My colleague Jenny Cheshire, at Queen Mary University of London, has called this variety Multicultural London English (usually abbreviated to MLE).[6] Cheshire and her co-researchers Sue Fox and Paul Kerswill have shown that MLE is a grammatically rich, systematic, and evolving system, quite distinct in many ways from standard English, though borrowing heavily from it. The team have been recording London teenagers over a number of years, transcribing and analysing their spontaneous speech, and have built up a large, and growing, corpus of examples.

We briefly came across relative clauses in Chapter 8. A relative clause is a sentence used to modify a noun or noun phrase. Here's an example from the Multicultural London English corpus. I've punctuated these to help understanding, though the corpus examples are presented without punctuation:

*Apparently, a chav is, like, someone **that wears like big gold chains**.*

The relative clause is in bold font. It's used to tell us something about the noun *someone*. You can see that this relative clause starts with the word *that*. But in English, and in MLE, it's possible to begin relative clauses in other ways. Here's another example from the corpus:

*Have you seen that protein drink **you can get**, like?*

The relative clause *you can get* is telling us something about the noun phrase *protein drink*. It doesn't start with the word *that*. There's no special grammatical marking that shows it's a relative clause except that the verb *get* is missing its Object, which is understood as the protein drink.

There's yet another way to make a relative clause, acceptable in both standard English and MLE. This uses a word like *who*, or *which*:

*I'm the only one **who's gone to college**.*

Finally, there's a way of making relative clauses that's stereotypically associated with the Cockney dialect of East London. There are very few examples of this in the corpus. This is one of these. Some information has been removed, as is standard, to preserve anonymity of the speakers:

No, that song **what they done at the E2 party in** **(name of** **street)**.

This example begins with the word *what*, as opposed to *that*, or *which*.

The team extracted almost 2,000 relative clauses from the corpus and analysed these. They split these examples into two groups depending on where the speakers were from. One group were relative clauses said by teenagers from inner London, in the borough of Hackney. The other was from an outer city borough, the far less multicultural Havering. It turned out that all of these teenagers mostly used the first three possibilities, and they used them, on the surface, quite similarly. But when Cheshire dug deeper into the data, it turned out that MLE was developing a new grammatical rule, found only in the inner city multilingual speakers from Hackney.

This new rule was embryonic. It showed as tendencies in the data, rather than as something that always applied. What was this tendency? The Hackney teenagers were favouring using *who* when the noun they were talking about was important enough to be referred to over and over again in subsequent sentences.

In linguistics, nouns and noun phrases that have this property are called Topics. Topics are marked using grammatical properties in many languages, though not in English. For example, in Japanese, the little particle *wa* appears after the Topic:

John   wa   gakusei   desu.
John   TOPIC   student   is

This sentence means something like, 'Talking about John, he's a student.' Japanese has a different particle *ga*, which is closer to what we've been calling a Subject in this book. So the following means 'John is a student':

John	*ga*	*gakusei*	*desu.*
John	SUBJECT	student	is

Some languages, like American Sign Language, use a special position in the sentence for Topics. Other languages, like Lakhota, a Siouan language spoken in North and South Dakota, have a special word for 'the' that they use for referring back to Topics. The Quechua languages, spoken in the Central Andes parts of Peru have very rich marking of topics, using a little grammatical word, -*qa*—the *q* here is pronounced not like English *qu*, but more like a *k* that is said very far back in the throat. The following examples are from Ayacucho Quechua:

kay-mi	*qiqa*
this	chalk

kay-qa	*qiqa-m*
this	chalk

Both sentences have a little affix, *mi*, shortened to just *m* in the second example. This expresses Evidentiality, which we met in Chapter 5 when we discussed Cherokee and Korean. But we're more interested in the word *qa*, which Quechua uses to mark a Topic. The first sentence lacks *qa* and just means 'This is a piece of chalk', but the second sentence uses *qa* to mark the Topic, the word *kay*, meaning 'this', and is closer to the English 'As for this, it's a piece of chalk'.

Topic is one of the grammatical meanings that Universal Grammar makes available for syntax, and some languages use particular grammatical words to signal this.

Topics are not usually marked in this way in English. English speakers do, occasionally, say things like *As for Anita, she's a great gardener*, marking out the Topic. Generally, though, Topic is not one of the features of Universal Grammar that is co-opted by the syntax of English.

To the MLE team's surprise, the teenagers in Hackney did co-opt Topic into the way they spoke English. When they used a relative clause that started with *who*, they were much more likely to treat the noun modified by that relative clause as a Topic. Cheshire looked at all the examples in the corpus of data where a noun was modified by a relative clause beginning with *who*, *that*, *what*, or nothing. She then examined the ten sentences that followed each of these, and counted how many times, in those ten sentences, the noun was referred back to.

Here's an example from the corpus, where one of our teenage speakers of MLE from Hackney is talking about his family. Each time there's a new clause, I've placed it on a new line:

> . . . my medium brother *who* moved to Antigua
> cos he's got a spinal disorder
> so he grows kinda slow
> so he is kinda short.
> People were swinging him about in my area.
> I thought 'what?'
> Now I lived near him then in north one.
> I still had my house in East London
> cos that's where I've lived born and raised
> like I had a house in East London
> where my nan lives

The relative clause is *who moved to Antigua*, and it tells us something about one of the speaker's brothers. Cheshire counted how many times that brother was referred back to in the next ten sentences. It turns out to be five of the ten. Compare this to the next example, where the speaker is talking about a terrorist attack on a bus in

London. The relative clause is *that were on the bus*, and it's telling us about a certain group of people:

> people *that* were on the bus were all different types of people
> so therefore all different types of families got affected by the same thing
> so natural and national disasters that happen in Britain
> everyone feels it
> and sometimes I think like religiously speaking
> sometimes I think.
> like them things should happen
> but there is still a benefit from like disasters
> because people do come together
> and we realize that people do get affected
> so yeah so I'm definitely a Londoner

In the ten sentences following the relative clause, there is no referencing back at all to the noun *people*. Cheshire went through the entire corpus, looking at all of the relative clauses that modify human nouns, and counted how often that noun was referred back to in the following ten sentences. She found a clear signal in the data. If the relative clause began with *who*, its noun was referred to more often. The word *who* was acting as a grammatical marker that the noun was a Topic.

We can see in this example how features from Universal Grammar get brought into a language through the social situation of the speakers. In the case of MLE, the English that the children are exposed to is more fragmentary than usual. Children whose parents are monolingual in English get a great deal of input in the language when they are young. The parents of the MLE children, however, generally have little English. This means that these children don't hear enough data to learn the standard English rule:

you can use either *who* or *that* in a relative clause if the noun is used to refer to a human; the notion of Topic isn't relevant.

All children come equipped by Universal Grammar with the idea of Topic, but children exposed to standard English see examples with *who* and with *that* both used with Topics and non-Topics. They don't associate the difference between the two words with that grammatical property. But the MLE children don't have enough evidence to know it's not at play in the language. Because of this, some MLE children come to associate the Topic property with *who*. We don't know exactly why this is, but here's a plausible idea. *Who* is only used in relative clauses that tell us something about humans. It's weird in standard English, and equally weird in MLE, to say something like *the chair who I sat on*. The word *who*, then, has a stronger connection to humans than the word *that*, which can be used in relative clauses that appear with nouns like *chair*. Now, humans are generally important players in conversations, and so are often Topics. That opens up the possibility that the MLE speakers unconsciously notice a relation between *who* and the grammatical notion of Topic, and, equally unconsciously, encode it as a grammatical rule. This scenario explains why it is *who* that ends up being used by the Hackney MLE speakers to mark that a noun phrase is a Topic.

Just as we saw with Burnouts and Jocks, individuals who feel a close social bond begin to speak in similar ways. Once some speakers begin to use *who* to mark out Topics, others begin to follow their lead, and the new rule spreads out through the community of speakers. The complex social situation of growing up in multilingual inner-city London, combined with the properties that Universal Grammar makes available, leads to a different grammar, the birth of a new way of speaking.

The evidence that the linguistic background of the speakers matters comes from another finding from the same study. The *who*-Topic pattern is only found in Hackney, the multicultural inner London borough. It wasn't found in Havering. Teenagers in Havering don't speak MLE. Their parents are, for the most part, native speakers of English, and the patterns of use that they have are like the patterns of use that we find more generally in English speakers. It's the Hackney teenagers, with their multilingual backgrounds, that innovate this new rule. They unconsciously organize their experiences, less English input, but rich communication with friends from all sorts of linguistic backgrounds, into a new mental grammar, which spreads through their community.

This last chapter has taken us in a different direction from the previous nine. It has been about those aspects of language that Merge does not touch. How is language used to express our social selves? Merge operates on discrete grammatical units, combining these endlessly to express linguistic meaning. But there is more to language, as we usually understand it, than this. Each sound we make, each choice of word, or inflection, can convey deep aspects of ourselves: who we feel we are, how we feel about what we are saying, what we want to accomplish with our words. Merge is the engine of the never-ending creativity of language, but we can also vary the sentences we create in myriad ways, expressing our identities, our emotions, and our intentions. One of the most exciting areas, for me, in current linguistics seeks to understand the relationships between linguistic structure, built by Merge, and language use.

GRAMMAR AND CULTURE · 251

I began this book by showing you the amazing creativity of human language. Our creative use of language is almost paradoxical. We are finite creatures, with only a few brief years' exposure to our languages while we are children, but we come to be able to use a system that is seemingly limitless in scope. I've argued that the way to think about this capacity of ours is to see that, at its heart, lies an incredibly simple piece of mental technology: Merge. Because Merge recursively builds hierarchies, with each application connecting to both meaning and sound, governed by Universal Laws of Language, there is no end to the complexity of the meaningful structures it builds. Merge gives us the ability to build the new worlds of ideas that have been so central to the successes and disappointments of our species. It makes language unlimited.

NOTES

CHAPTER 1

1. Darwin, Charles. 1871. *The Descent of Man, and Selection in Relation to Sex.* London: Murray.
2. Goldberg, Adele. 1995. *Constructions.* Chicago: Chicago University Press.
   Tomasello, Michael. 2005. *Constructing a Language.* Cambridge, MA: Harvard University Press.
3. Pearl, Lisa and Jon Sprouse. 2013. Computational models of acquisition for islands. In Jon Sprouse and Norbert Hornstein (eds), *Experimental Syntax and Island Effects.* Cambridge: Cambridge University Press. 109–31.
4. Kandybowicz, Jason. 2006. Comp-trace effects explained away. Proceedings of the 25th West Coast Conference on Formal Linguistics. Somerville, MA: Cascadilla Proceedings Project.
5. Pesetsky, David. 2017. Complementizer-Trace Effects. *The Wiley Blackwell Companion to Syntax.* Oxford: Wiley 1–34.

CHAPTER 2

1. McCulloch, Gretchen. 2019. *Because Internet: Understanding the New Rules of Language.* New York: Riverhead Press.
2. Twemoji graphics made by Twitter and other contributors, licensed under CC-BY 4.0: https://creativecommons.org/licenses/by/4.0/
3. Trauth, K.M., Hora, S.C., and Guzowski, R.V. 1993. *Expert judgment on markers to deter inadvertent human intrusion into the Waste Isolation Pilot Plant.* United States. Web doi:10.2172/10117359.
4. Balasuriya, Lakshika, Sanjaya Wijeratne, Derek Doran, and Amit Sheth. 2016. Signals Revealing Street Gang Members on Twitter. In *Workshop on Computational Approaches to Social Modeling,* Volume 4. Bellevue, WA, USA.

5. Adger, David, Daniel Harbour, and Laurel Watkins. 2009. *Mirrors and Microparameters: from Phrase Structure to Free Word Order.* Cambridge: Cambridge University Press.
6. Shannon, Claude. 1948. A Mathematical Theory of Communication. *Bell System Technical Journal* 27: 379–423.

CHAPTER 3

1. See www.bostoncriminallawyersblog.com/massachusetts-drug-violation-applies-public-playgrounds accessed 8 February 2019.
2. Dennett, Daniel. 1978. *Brainstorms,* Cambridge, MA: MIT Press.
3. Bock, Kathryn. 1986. Syntactic persistence in language production. *Cognitive Psychology* 18.3: 355–87.
4. Pallier, Christophe, Anne-Dominque Devauchelle, and Stanslaus Dehaene. 2011. Cortical representation of the constituent structure of sentences. *Proceedings of the National Academy of Sciences* 108: 2522–7.
5. Ding, Nai, Lucia Melloni, Hang Zhang, Xing Tian, and David Poeppel. 2016. Cortical tracking of hierarchical linguistic structures in connected speech. *Nature Neuroscience* 19: 158–64.
6. Chomsky, Noam. 1986. *Knowledge of Language: Its Nature, Origin, and Use.* New York: Praeger.

CHAPTER 4

1. Goldin-Meadow, Susan. 2005. *The Resilience of Language.* New York: Psychology Press.
2. Goldin-Meadow, Susan and Charles Yang. 2016. Statistical evidence that a child can create a combinatorial linguistic system without external linguistic input: Implications for language evolution. *Neuroscience and Biobehavioral Reviews* 81: 150–7.
3. Haviland, John B. 2016. "But you said 'four sheep'!": (sign) language, ideology, and self (esteem) across generations in a Mayan family. *Language and Communication* 46: 62–94.
Haviland, John B. 2013. The emerging grammar of nouns in a first

generation sign language: Specification, iconicity, and syntax. *Gesture* 13: 309–53.

4. Wojdak, Rachel. 2000. Nuu-chah-nulth modification: Syntactic evidence against category neutrality. Papers for the 35th International Conference on Salish and Neighbouring Languages. UBCWPL. Vol. 3.

5. Haviland uses the term 'specifiers' in his work, partly because the term 'classifier' has a special technical meaning in sign language research that is different from the way that the term is used in work on spoken languages. When researchers look cross-linguistically at noun syntax, they tend to use the term 'classifier' as I do here. But beware if you see it in sign language work, where it will likely have a different meaning.

6. Drawings in Figure 1 and Figure 2 based on Haviland (2016).

7. Senghas, Ann, Sotaro Kita, and Asli Özyürek. 2004. Children Creating Core Properties of Language: Evidence from an Emerging Sign Language in Nicaragua. *Science* 305: 1779.

8. Shultz, Sarah, Athena Vouloumanos, Randi H. Bennett, and Kevin Pelphrey. 2014. Neural specialization for speech in the first months of life. *Developmental Science* 17: 766–74.

9. Castro, Mario, Rodolfo Cuerno, Matteo Nicoli, Luis Vázquez, and Josephus G. Buijnsters. 2012. Universality of cauliflower-like fronts: from nanoscale thin films to macroscopic plants. *New Journal of Physics* 14.103039.

10. Joos, Martin. 1957. Readings in Linguistics 1. Chicago: University of Chicago Press.

11. Evans, Nicholas and Stephen C. Levinson. 2009. The myth of language universals: Language diversity and its importance for cognitive science. *Behavioral and Brain Sciences* 32: 429–48.

CHAPTER 5

1. A great read on how languages are constructed is Peterson, David. 2015. *The Art of Language Invention*. New York: Penguin.

2. Harbour, Daniel. 2014. Paucity, abundance, and the theory of number. *Language* 90: 185–229.
3. Adger, David. 2015. Syntax. *WIREs Cognitive Science* 6: 131–47.
4. Musso, Maria-Cristina, Andrea Moro, Volkmar Glauche, Michel Rijntjes, Jürgen Reichenbach, Christian Büchel, and Cornelius Weiller. 2003. Broca's area and the language instinct. *Nature Neuroscience* 6: 774.
5. Rett, Jessica and Sarah E. Murray. 2013. A semantic account of Mirative evidentials. Proceedings of SALT 23: 453–72, Cornell University.
6. The *ka* suffix that appears on *Toli* marks that word is a Subject in the sentence.
7. Thoms, Gary, David Adger, Caroline Heycock, and Jennifer Smith. 2019. Syntactic variation and auxiliary contraction: the surprising case of Scots. *Language*. To appear.

CHAPTER 6

1. The spectrogram was created using the software program Praat www.fon.hum.uva.nl/praat
2. Liberman, Alvin.M., Katherine Safford Harris, Howard S. Hoffman, and Belver C. Griffith. 1957. The discrimination of speech sounds within and across phoneme boundaries. *Journal of Experimental Psychology* 54: 358–68.
3. Eimas Peter D., Joanne L. Miller, and Peter W. Jusczyk. 1987. On infant speech perception and the acquisition of language. In Steven Harnad (ed.), *Categorical Perception*. Cambridge: Cambridge University Press. 161–95.
4. Kuhl, Patricia K. and James D. Miller. 1975. Speech perception by the chinchilla: voice-voiceless distinction in alveolar plosive consonants. *Science* 190: 69–72.
5. Ghislaine Dehaene-Lambertz. 2017. The human infant brain: A neural architecture able to learn language. *Psychonomic Bulletin and Review* 24: 48–55.
6. Brent, Michael R. and Jeffrey Mark Siskind. 2001. The role of exposure to isolated words in early vocabulary development. *Cognition* 81: B33-B44.

7. Saffran, Jenny R., Richard N. Aslin, and Elissa L. Newport. 1996. Statistical learning by 8-month-old infants. *Science* 274: 1926–8.

8. Hauser, Marc D., Elissa L. Newport, and Richard N. Aslin. 2001. Segmentation of the speech stream in a non-human primate: statistical learning in cotton-top tamarins. *Cognition* 78: B53–B64.

9. Newport, Elissa, Marc Hauser, Geertrui Spaepen, and Richard Aslin. 2004. Learning at a distance II. Statistical learning of non-adjacent dependencies in a non-human primate. *Cognitive Psychology* 49: 85–117.

10. Pons, Ferran, and Juan M. Toro. 2010. Structural generalizations over consonants and vowels in 11-month-old infants. *Cognition* 116: 361–7.

    Mora, Daniela de la and Juan M. Toro. 2013. Rule learning over consonants and vowels in a non-human animal. *Cognition* 126: 307–12.

11. Johnson, Elizabeth K. and Peter W. Jusczyk. 2001. Word segmentation by 8-month-olds: When speech cues count more than statistics. *Journal of Memory and Language* 44: 548–67.

    Thiessen, Erik D. and Jenny R. Saffran. 2003. When cues collide: use of stress and statistical cues to word boundaries by 7- to 9-month-old infants. *Developmental Psychology* 39: 706–16.

    Börschinger, Benjamin and Mark Johnson. 2014. Exploring the Role of Stress in Bayesian Word Segmentation using Adaptor Grammars. *Transactions of the Association of Computational Linguistics*, 2: 93–104.

12. Jusczyk, Peter. 1999. How infants begin to extract words from speech. *Trends in Cognitive Sciences* 3: 323–8.

13. Quine, W. V. O. 1960. *Word and Object*. New York: John Wiley and Sons.

14. Spelke, Elizabeth S. and Katherine D. Kinzler. 2007. Core knowledge. *Developmental Science* 10: 89–96.

15. Kaminski, Juliane, Josep Call, and Julia Fischer. 2004. Word learning in a domestic dog: evidence for "fast mapping". *Science* 304: 1682–3.

16. Seyfarth, Robert M., Dorothy L. Cheney, and Peter Marler. 1980. Vervet monkey alarm calls: semantic communication in a free-ranging primate. *Animal Behaviour* 28: 1070–94.

17. Zuberbühler, Klaus. 2015. Linguistic capacity of non-human animals. Wiley Interdisciplinary Reviews. *Cognitive Science* 6: 313–21.

CHAPTER 7

1. Savage-Rumbaugh, Susan. 1986. *Ape Language: From Conditioned Response to Symbol.* New York: Columbia University Press.
2. Truswell, Robert. 2017. Dendrophobia in bonobo comprehension of spoken English. *Mind and Language* 32: 395–415.
3. Bybee, Joan. 2010. *Language, Usage and Cognition.* Cambridge: Cambridge University Press.
4. Adger, David. 2003. *Core Syntax: A Minimalist Approach.* Oxford: Oxford University Press.
5. The Binding Theory contains a number of Laws that collectively constrain how various kinds of words can be linked in meaning in sentences. The Pronoun-Name Law is more usually called Principle C of the Binding Theory. See Büring, Daniel. 2005. *Binding Theory.* Cambridge: Cambridge University Press.
6. Bruening, Benjamin. 2001. Constraints on dependencies in Passamaquoddy. Algonquian Papers Archive 32.
7. Andrews, Avery. 2017. Prenominal Possessives in English: What does the Stimulus Look Like? Lingbuzz: https://ling.auf.net/lingbuzz/003568
8. Ambridge, Ben and Elena Lieven. 2015. A Constructivist Account of Child Language Acquisition. *The Handbook of Language Emergence* 87: 478.
9. Valian, Virginia, Stephanie Solt, and John Stewart. 2009. Abstract categories or limited-scope formulae? The case of children's determiners. *Journal of Child Language* 36: 743–78.
10. Sutton, Megan, Michael Fetters, and Jeffrey Lidz. 2012. Parsing for Principle C at 30 months. *Proceedings of the 36th Boston University Conference on Language Development.* Somerville: Cascadilla Press.

CHAPTER 8

1. Lewis, Mike, Dennis Yarats, Yann Dauphin, Devi Parikh, and Dhruv Batra. 2017. Deal or no deal? End-to-end learning for negotiation

dialogues. *Proceedings of the 2017 Conference on Empirical Methods in Natural Language Processing*. Copenhagen: Association for Computational Linguistics. 2443–53.

2. Winograd, Terry. 1972. Understanding natural language. *Cognitive Psychology* 3: 1–191.

3. Bender, D. 2015. Establishing a Human Baseline for the Winograd Schema Challenge. In M. Glass and J-H. Kim (eds), *Proceedings of the 26th Modern Artificial Intelligence and Cognitive Science conference*. University of Cincinnati, 39–45.

4. Marcus, Mitchell, Mary Ann Marcinkiewicz, and Beatrice Santorini. 1993. Building a large annotated corpus of English: The Penn Treebank. *Computational Linguistics* 19: 313–30.

5. https://books.google.com/ngrams

6. Miller, George A. and Jennifer A. Selfridge. 1950. Verbal context and the recall of meaningful material. *The American Journal of Psychology* 63: 176–85.

7. Linzen, Tal, Emmanuel Dupoux, and Yoav Goldberg. 2016. Assessing the ability of LSTMs to learn syntax-sensitive dependencies. arXiv preprint arXiv:1611.01368.

CHAPTER 9

1. Mandelbrot, Benoit B. 1982. *The Fractal Geometry of Nature*. New York: WH Freeman.

2. Chomsky, Noam. 1995. *The Minimalist Program*. Cambridge, MA: MIT Press.

3. Derbyshire, Desmond. 1977. Word order universals and the existence of OVS languages. *Linguistic Inquiry* 8: 590–9.

4. Hauser, Marc D., Noam Chomsky, and W. Tecumseh Fitch. 2002. The faculty of language: What is it, who has it, and how did it evolve? *Science* 298: 1569–79.

5. Chomsky, Noam. 2014. Minimal recursion: exploring the prospects. In Tom Roeper and Margaret Speas (eds) *Recursion: Complexity in Cognition*. Cham: Springer. 1–15.

6. The basic idea is in Chomsky, Noam. 1995. *The Minimalist Program*. Cambridge, MA: MIT Press, chapter 3.

See also the appendix to Fox, Danny and David Pesetsky. 2005. Cyclic linearization of syntactic structure. *Theoretical Linguistics* 31: 1–45.

7. Adger, David. 2010. Gaelic Syntax. In Murray Watson and Michelle Macleod (eds), *The Edinburgh Companion to the Gaelic Language*. Edinburgh: Edinburgh University Press. 304–51.

8. If I'm being totally honest, I should say that the approach I've presented here is the consensus approach to Verb Subject Object order. I myself have suggested something different, but that would take the discussion into the knotty details of syntactic theory. See Adger, David. 2013. *A Syntax of Substance*. Cambridge, MA: MIT Press.

9. Ritter, Elizabeth and Martina Wiltschko. 2009. Varieties of INFL: Tense, location, and person. In Jeroen van Craenenbroek (ed.), *Alternatives to Cartography*. Berlin: de Gruyter Mouton. 153–202.

CHAPTER 10

1. Eckert, Penelope. 2000. *Linguistic Variation as Social Practice*. Oxford: Blackwell.

2. Munson, Ben, Elizabeth McDonald, Nancy DeBoe, and Aubrey White. 2006. The acoustic and perceptual bases of judgments of women and men's sexual orientation from read speech. *Journal of Phonetics* 34: 202–40.

3. Podesva, Robert J. and Janneke Van Hofwegen. 2016. /s/exuality in smalltown California: Gender normativity and the acoustic realization of /s/. In Erez Levon and Ronald Beline Mendes (eds) *Language, Sexuality, and Power: Studies in Intersectional Sociolinguistics*. Oxford: Oxford University Press. 168–88.

4. Smith, Jennifer and Mercedes Durham. 2019. *Sociolinguistic Variation in Children's Language*. Cambridge: Cambridge University Press. The original version of this conversation is on pages 144–5.

5. Potts, Chris. 2007. The Expressive Dimension. *Theoretical Linguistics* 33: 165–98.

6. Cheshire, Jenny, David Adger, and Sue Fox. 2013. Relative *who* and the actuation problem. *Lingua* 126: 51–77.

ACKNOWLEDGMENTS

This book has been a long time in the writing. I'd like to thank my editor at OUP, Julia Steer, for encouragement, and for finding a title that helped me focus my writing. Many thanks too to the five reviewers of the original proposal for being critical, but extremely helpful; to Adam Schembri, Michelle Sheehan, and Rob Truswell for detailed comments on an early draft; thanks too to Erez Levon, Robyn Orfitelli, Dina Rider, Devyani Sharma, Jennifer Smith, and Peter Tallack for useful discussions; and to Adam Chong and Satoshi Tomioka for data help; special thanks to Ronnie Kroon for being the best non-linguist reader I could have hoped for.

For keeping me fuelled with coffee and cocktails, thanks to Anders, Karen, and Hedwig at *All you Read is Love*, who were there at the beginning of this book. Sorry I took so long to write this that the cafe closed before I finished it! For giving me a precious clear last few days in their Narrowsburg house, where I finally brought the last chapter to a close, many thanks to Aaron, Ilya, and Pip.

In my first book, published about 15 years ago, I ended the acknowledgments with thanks to Anson, *sine quō nihil*. Things haven't changed.

INDEX